Problem Solvers

Edited by L. Marder

Senior Lecturer in Mathematics, University of Southampton

No. 16

Groups

Problem Solvers

Groups

D. A. R. WALLACE

Professor of Mathematics, University of Stirling

LONDON . GEORGE ALLEN & UNWIN LTD

RUSKIN HOUSE   MUSEUM STREET

First published 1974

© George Allen & Unwin Ltd, 1974

ISBN 0 04 519012 7 *hardback*
     0 04 519013 5 *paperback*

Set in 10 on 12 pt 'Monophoto' Times Mathematics Series 569
Printed in Great Britain by Page Bros (Norwich) Ltd., Norwich

Contents

Chapter 1

Sets, Mappings, Relations

1.1 In everyday parlance the words 'assembly', 'class', 'group', 'set', etc., are used indiscriminately to denote collections of similar objects. Mathematicians use the single word *set* to denote an arbitrary collection and the word *element* for a member of the collection or set. In this book, capital letters will be used to denote sets and small letters to denote elements. To indicate that the symbol a is an element of the set A, we write $a \in A$ and read this as 'a is an element of A'. Thus if A consists of the integers 4, 5, 6, 7, 8 we write $4 \in A$, and to indicate that A consists of these numbers we use curly brackets and write $A = \{4, 5, 6, 7, 8\}$. In general if the set A has a finite number n (say) of distinct elements $a_1, a_2, \ldots, a_n$ we write

$$A = \{a_1, a_2, \ldots, a_n\}. \tag{1.1}$$

Notice that as we are interested only in the totality of the elements in A, the order of these elements inside the curly brackets is irrelevant, thus

$$\{4, 5, 6, 7, 8\} = \{5, 7, 4, 8, 6\} = \{8, 7, 6, 4, 5\} = \ldots \tag{1.2}$$

Any subcollection B of r elements ($r \geqslant 1$) of the set $A = \{a_1, a_2, \ldots, a_n\}$ is of the form

$$B = \{a_{i_1}, a_{i_2}, \ldots, a_{i_r}\}$$

where the a_{i_j} are distinct and each a_{i_j} is an a_t for some t ($1 \leqslant t \leqslant n$). We call such a subcollection B a *subset* of A. The sets $\{4, 5, 6, 7\}$, $\{4\}$, $\{5, 6\}$ are all subsets of $\{4, 5, 6, 7\}$.

Problem 1.1 The set A consists of the elements a, b, c, d (assumed to be distinct). Write down all subsets of A having at least one element.

Solution. First we observe that A is a subset of A. There are four subsets of A having three elements each, namely $\{a, b, c\}$, $\{a, b, d\}$, $\{a, c, d\}$, $\{b, c, d\}$. There are six subsets of A having two elements each, namely $\{a, b\}$, $\{a, c\}$, $\{a, d\}$, $\{b, c\}$, $\{b, d\}$, $\{c, d\}$ and four subsets having one element each, namely $\{a\}$, $\{b\}$, $\{c\}$, $\{d\}$. $\qquad\square$

It is convenient to consider, as a subset of a set A, the set consisting of no elements whatever; this set is called the *empty set* and is denoted by $\varnothing$ (a letter in the Scandanavian alphabet), other subsets of A are called *non-empty*. Thus the set $\{1, 2, 3\}$ has the eight subsets $\{1, 2, 3\}$, $\{1, 2\}$, $\{1, 3\}$, $\{2, 3\}$, $\{1\}$, $\{2\}$, $\{3\}$, $\varnothing$; notice that we distinguish between the element 3 (say) and the set $\{3\}$ consisting of the single element 3.

Problem 1.2 Let A be the set having n distinct elements $a_1, a_2, \ldots, a_n$. Prove that A has exactly 2^n distinct subsets.

Solution. We have $A = \{a_1, a_2, \ldots, a_n\}$. Any nonempty subset of A consisting of r distinct elements $(r \geqslant 1)$ is of the form $\{a_{i_1}, a_{i_2}, \ldots, a_{i_r}\}$ where the a_{i_j} are distinct and where the order of the a_{i_j} is irrelevant. Thus the number of subsets of r elements is the number of ways of selecting r elements from n elements, this number is the binomial coefficient $\binom{n}{r}$. Hence the total number of subsets, including $\varnothing$, is $1 + \sum_{r=1}^{n} \binom{n}{r}$ and, by the binomial theorem, we have $1 + \sum_{r=1}^{n} \binom{n}{r} = (1+1)^n = 2^n$. $\qquad\square$

Problem 1.3 How many subsets can be formed from the letters of the word 'mathematics'?

Solution. Some of the letters are repeated and so the set A consisting of the letters of 'mathematics' is $A = \{m, a, t, h, e, i, c, s\}$. We see that A has 8 elements and so A has $2^8 = 256$ subsets. $\qquad\square$

So far we have mentioned only finite sets whereas many of the sets to be encountered in mathematics possess an infinity of elements. Thus the set $\mathbb{Z}$ of all integers is infinite and we write

$$\mathbb{Z} = \{0, \pm 1, \pm 2, \ldots\}, \qquad (1.3)$$

the dots indicating that the integers are continued in the obvious manner ['$\mathbb{Z}$' for Zahl (German) = number]. The use of dots is not always possible and a set is often specified by some property P (say); we write

$$\{x : x \text{ has property } P\} \qquad (1.4)$$

to denote the set consisting precisely of those x having the given property P. In this notation we have

$$\mathbb{Z} = \{x : x \text{ is an integer}\} = \{x : x = 0, \pm 1, \pm 2, \ldots\}. \qquad (1.5)$$

The set of even integers between 1 and 10 inclusive is

$$\{n : n \in \mathbb{Z}, \ 1 \leqslant n \leqslant 10, \ 2 \text{ divides } n\}. \qquad (1.6)$$

Problem 1.4 Prove that the set $A = \{n : n \in \mathbb{Z}, \ n^2 \leqslant 9\}$ is finite.

Solution. A is the set of integers each of which has square at most 9. Thus $A = \{-3, -2, -1, 0, 1, 2, 3\}$ which is finite. $\qquad\square$

If X is a subset of a set Y we write $X \subseteq Y$ and read this as 'X is contained in or equal to Y', thus $X \subseteq Y$ if and only if $x \in X$ implies that $x \in Y$. X is a *proper* subset of Y if $X \subseteq Y$ and $X \neq Y$. Two sets X and Y are equal if and only if they have exactly the same elements; consequently

$X = Y$ if and only if $X \subseteq Y$ and $Y \subseteq X$. If x is not an element of the set A we write $x \notin A$ and if X is not a subset of the set A we write $X \nsubseteq A$. Thus $3 \in \mathbb{Z}$ but $\frac{1}{2} \notin \mathbb{Z}$ and $\{a, i, u\} \subseteq \{a, e, i, o, u\}$ but $\{a, b\} \nsubseteq \{a, e, i, o, u\}$.

Numbers of the form $\dfrac{m}{n}$ ($m, n \in \mathbb{Z}$, $n \neq 0$) are called *rational numbers*, the set of rational numbers being denoted by $\mathbb{Q}$ ('$\mathbb{Q}$' for 'quotient'), thus

$$\mathbb{Q} = \left\{ x : x = \frac{m}{n}; \quad m, n \in \mathbb{Z}, \quad n \neq 0 \right\}. \tag{1.7}$$

The numbers $\frac{1}{2}, \frac{3}{5}, \frac{7}{19}, \dfrac{m}{1}$ are elements of $\mathbb{Q}$ which clearly contains $\mathbb{Z}$ as a proper subset.

Problem 1.5 What is $\{x : x \in \mathbb{Q}, \quad x^2 = 2\}$?

Solution. It is known that $\sqrt{2}$ is not a rational number and so the above set has no elements, in other words $\{x : x \in \mathbb{Q}, \quad x^2 = 2\} = \varnothing$. $\square$

Problem 1.6 Let A be the set of integers from 1 to 7 inclusive and let B be the set of even integers between 1 and 13. Write down A and B in set-theoretical notation. Let X be the set of integers common to both A and B and let Y be the set of integers in either A or B. Find X and Y.

Solution. $A = \{n : n \in \mathbb{Z}, \quad 1 \leqslant n \leqslant 7\}$
$\qquad\quad = \{1, 2, 3, 4, 5, 6, 7\}$
$\qquad B = \{x : x = 2n, \quad n \in \mathbb{Z}, \quad 1 \leqslant x \leqslant 13\}$
$\qquad\quad = \{2, 4, 6, 8, 10, 12\},$
$\quad X = \{2, 4, 6\}, \qquad Y = \{1, 2, 3, 4, 5, 6, 7, 8, 10, 12\}.$ $\square$

The set of elements common to two given sets A and B is called the *intersection* of A and B, written $A \cap B$. Clearly $A \cap B = B \cap A$ and $A \cap A = A$. The set of elements in either of two given sets A and B is called the *union* of A and B, written $A \cup B$. Clearly $A \cup B = B \cup A$ and $A \cup A = A$.

Let A, B, X be sets such that $A \subseteq B$. From the definitions we have $A \cap X \subseteq B \cap X \subseteq B$ and $A \subseteq A \cup X \subseteq B \cup X$.

Problem 1.7 Let A, B be sets. Prove that $A \subseteq B$ if and only if $A \cap B = A$.

Solution. If $A \subseteq B$ we have $A = A \cap A \subseteq A \cap B \subseteq A$ and thus $A = A \cap B$. Conversely if $A \cap B = A$ we have $A = A \cap B \subseteq B$.

Similarly we may show that $A \subseteq B$ if and only if $A \cup B = B$. $\square$

Problem 1.8 Let $A = \{1, 3, 5, 7, 9, 11\}$, $B = \{1, 5, 8\}$ and $C = \{12, 13\}$. Write down $A \cap B$, $B \cap C$, $(A \cap B) \cap C$, $A \cap (B \cap C)$, $A \cup B$, $B \cup C$, $(A \cup B) \cup C$, $A \cup (B \cup C)$.

Solution. $A \cap B = \{1, 5\}$, $B \cap C = \emptyset$, $(A \cap B) \cap C = \emptyset = A \cap (B \cap C)$,
$A \cup B = \{1, 3, 5, 7, 8, 9, 11\}$, $B \cup C = \{1, 5, 8, 12, 13\}$,
$(A \cup B) \cup C = \{1, 3, 5, 7, 8, 9, 11, 12, 13\} = A \cup (B \cup C)$.

We note that $(A \cap B) \cap C$ is the set of elements common to A, B, C; we omit the bracketing and write this set as $A \cap B \cap C$. Similarly we write $A \cup B \cup C$ for $(A \cup B) \cup C$. $\qquad\qquad\qquad\qquad\qquad\square$

Let $A_1, A_2, \ldots, A_n$ be n sets. The *intersection* of these n sets is defined to be the set of elements common to all of the A_i and is denoted by $A_1 \cap A_2 \cap \ldots \cap A_n$ or by $\bigcap_{j=1}^{n} A_j$. The *union* of these n sets is defined to be the set of elements in at least one of the A_i and is denoted by $A_1 \cup A_2 \cup \ldots \cup A_n$ or by $\bigcup_{j=1}^{n} A_j$.

We can regard n sets $A_1, A_2, \ldots, A_n$ as being a family of sets indexed by the set $\{1, 2, \ldots, n\}$, i.e. for each $i \in \{1, 2, \ldots, n\}$ we have a set A_i. This is perhaps a tortuous description for a finite number of sets but the notion generalises to the case of a family of sets whose indexing set is an arbitrary set Λ (say), i.e. for each $\lambda \in \Lambda$ we have a set A_λ of the family. The *intersection* of the A_λ ($\lambda \in \Lambda$) is defined to be the set of elements common to all the A_λ and is denoted by $\bigcap_{\mu \in \Lambda} A_\mu$. In a corresponding manner we define the *union* of the A_λ ($\lambda \in \Lambda$) and denote it by $\bigcup_{\mu \in \Lambda} A_\mu$.

Problem 1.9 Let $A_1 = \{s, i, c\}$, $A_2 = \{t, r, a, n, s, i, t\}$, $A_3 = \{g, l, o, r, i, a\}$, $A_4 = \{m, u, n, d, i\}$. What are $A_1 \cap A_2 \cap A_3 \cap A_4$ and $A_1 \cup A_2 \cup A_3 \cup A_4$?

Solution. $A_1 \cap A_2 \cap A_3 \cap A_4 = \{i\}$,
$$A_1 \cup A_2 \cup A_3 \cup A_4 = \{a, c, d, g, i, l, m, n, o, r, s, t, u\}. \qquad\square$$

Problem 1.10 Let $\mathbb{R}$ denote the set of real numbers and let A, B, C be the subsets of $\mathbb{R}$ given by $A = \{x : x \in \mathbb{R}, \ 0 \leqslant x \leqslant 3\}$, $B = \{x : x \in \mathbb{R}, -1 \leqslant x \leqslant 2\}$, $C = \{x : x \in \mathbb{R}, \ -2 \leqslant x \leqslant 1\}$. Show that $A \cap (B \cup C) = (A \cap B) \cup (A \cap C)$ and that $A \cup (B \cap C) = (A \cup B) \cap (A \cup C)$.

Solution. Before proving the results we make two remarks. First, the notation $0 \leqslant x \leqslant 3$, by long-established usage, implies that x is a real number and so, without loss of clarity, we write

$$A = \{x : 0 \leqslant x \leqslant 3\}, \ B = \{x : -1 \leqslant x \leqslant 2\}, \ C = \{x : -2 \leqslant x \leqslant 1\}.$$

Second, A, B, C are, in fact, closed intervals on the real line. Figure 1.1 should assist in understanding the argument.

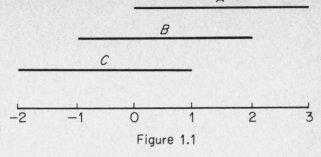

Figure 1.1

We have $B \cup C = \{x : -2 \leqslant x \leqslant 2\}$, $A \cap B = \{x : 0 \leqslant x \leqslant 2\}$, $A \cap C = \{x : 0 \leqslant x \leqslant 1\}$. Thus $A \cap (B \cup C) = \{x : 0 \leqslant x \leqslant 2\} = (A \cap B) \cup (A \cap C)$. Also we have $B \cap C = \{x : -1 \leqslant x \leqslant 1\}$, $A \cup B = \{x : -1 \leqslant x \leqslant 3\}$, $A \cup C = \{x : -2 \leqslant x \leqslant 3\}$. Thus $A \cup (B \cap C) = \{x : -1 \leqslant x \leqslant 3\} = (A \cup B) \cap (A \cup C)$. □

It can be established that for any three sets A, B, C the operations of $\cap$ and $\cup$ are related by the formulae:

$$A \cap (B \cup C) = (A \cap B) \cup (A \cap C) \tag{1.8}$$

$$A \cup (B \cap C) = (A \cup B) \cap (A \cup C). \tag{1.9}$$

These formulae admit of easy generalisations. Thus if $B_1, B_2, \ldots, B_n$ are n sets we have

$$A \cap \left(\bigcup_{j=1}^{n} B_i \right) = \bigcup_{j=1}^{n} (A \cap B_j) \tag{1.10}$$

$$A \cup \left(\bigcap_{j=1}^{n} B_i \right) = \bigcap_{j=1}^{n} (A \cup B_j) \tag{1.11}$$

and for a family of sets indexed by a set Λ with B_λ ($\lambda \in \Lambda$) as a typical set of the family

$$A \cap \left(\bigcup_{\mu \in \Lambda} B_\mu \right) = \bigcup_{\mu \in \Lambda} (A \cap B_\mu) \tag{1.12}$$

$$A \cup \left(\bigcup_{\mu \in \Lambda} B_\mu \right) = \bigcap_{\mu \in \Lambda} (A \cup B_\mu). \tag{1.13}$$

Problem 1.11 Prove formula 1.8.

Solution. We prove first that $A \cap (B \cup C) \subseteq (A \cap B) \cup (A \cap C)$. Let $x \in A \cap (B \cup C)$. Then $x \in A$ and $x \in B \cup C$. Since $x \in B \cup C$ either $x \in B$ or $x \in C$. If $x \in B$ then $x \in A$ and $x \in B$ and so $x \in A \cap B$. Hence $x \in (A \cap B) \cup (A \cap C)$. Similarly if $x \in C$ we deduce that $x \in (A \cap B) \cup (A \cap C)$ and so we have shown that $A \cap (B \cup C) \subseteq (A \cap B) \cup (A \cap C)$. We now prove

5

that $(A \cap B) \cup (A \cap C) \subseteq A \cap (B \cup C)$. Let $y \in (A \cap B) \cup (A \cap C)$. Then either $y \in A \cap B$ or $y \in A \cap C$. If $y \in A \cap B$ then $y \in A$ and $y \in B$, thus certainly $y \in A$ and $y \in B \cup C$, hence $y \in A \cap (B \cup C)$. Similarly if $y \in A \cap C$ we deduce that $y \in A \cap (B \cup C)$ and so we have shown that $(A \cap B) \cup (A \cap C) \subseteq A \cap (B \cup C)$. $\square$

Problem 1.12 Let A, X, Y be sets such that $A \cap X = A \cap Y$ and $A \cup X = A \cup Y$. Prove that $X = Y$.

Solution. Since $X \subseteq X \cup A$ we have, by Problem 1.7, that $X \cap (X \cup A) = X$ and that $Y \cap (Y \cup A) = Y$. Hence

$$X = X \cap (X \cup A) = X \cap (Y \cup A)$$
$$= (X \cap Y) \cup (X \cap A) \qquad \text{(by 1.8)}$$
$$= (X \cap Y) \cup (Y \cap A) = (Y \cap X) \cup (Y \cap A)$$
$$= Y \cap (Y \cup A) \qquad \text{(by 1.8)}$$
$$= Y. \qquad \square$$

We are accustomed to represent the points of the Euclidean plane as ordered pairs (x, y) of real numbers x and y, the ordering of the pair being vital as, for example, $(5, 8)$ and $(8, 5)$ are distinct points. In a similar manner, given two sets X and Y we form the set, denoted by $X \times Y$, of ordered pairs of elements of the form (x, y) $(x \in X, y \in Y)$, i.e.

$$X \times Y = \{(x, y) : x \in X, \quad y \in Y\}. \qquad (1.14)$$

The set $X \times Y$ is called the *Cartesian product* of X and Y (after R. Descartes, French, 1596–1650). The Euclidean plane, considered simply as a set of points, is then the set $\mathbb{R} \times \mathbb{R}$. The set of complex numbers, denoted by $\mathbb{C}$, is sometimes introduced as a set of ordered pairs of real numbers (with appropriate laws of addition, etc.) and so we may write $\mathbb{C} = \mathbb{R} \times \mathbb{R}$.

Problem 1.13 Let $X = \{1, 2\}$ and $Y = \{a, b\}$. Find $X \times Y$ and $Y \times X$.

Solution. $\qquad X \times Y = \{(1, a), (1, b), (2, a), (2, b)\}$
$$Y \times X = \{(a, 1), (a, 2), (b, 1), (b, 2)\}. \qquad \square$$

We extend the above notions to form the Cartesian product $X_1 \times X_2 \times \dots \times X_n$ of the n sets $X_1, X_2, \dots, X_n$ taken in order. Thus
$$X_1 \times X_2 \times \dots \times X_n = \{(x_1, x_2, \dots, x_n) : x_i \in X_i, \quad i = 1, 2, \dots, n\}. \quad (1.15)$$

Problem 1.14 Let X_i be a finite set having r_i elements $(i = 1, 2, \dots, n)$. Prove that $X_1 \times X_2 \times \dots \times X_n$ has $r_1 r_2 \dots r_n$ elements.

Solution. In the n-tuple $(x_1, x_2, \dots, x_n)$ there are r_1 possibilities for x_1, r_2 possibilities for x_2 and so on. These possibilities are independent and hence the result holds. $\square$

1.2 Mappings In mathematics we are often concerned with ways of associating one set with another; given $x \in \mathbb{R}$ we can form $\sin x$ and we know, from the properties of the sine function, that $\sin x$ may be any real number between -1 and $+1$ inclusive; in not too precise terms the sine function associates with $\mathbb{R}$ the set $\{y : -1 \leqslant y \leqslant 1\}$. In general let X and Y be two nonempty sets, if with each $x \in X$ we can associate, by some given means, a uniquely determined element $f(x) \in Y$ then we call f a *function* or *mapping* of X *into* Y. We use either of the following notations

$$f : X \rightarrow Y \tag{1.16}$$

$$X \xrightarrow{f} Y \tag{1.17}$$

to indicate that X, Y are nonempty sets and that f is a mapping from X into Y. We say that $f(x)$ is the *image* of x under f. X is called the *domain* of f and the subset of Y consisting of all elements of the form $f(x)$ is called the *range* of f or the *image* of f. We note the image under f by Im f or by $f(X)$, thus

$$f(X) = \text{Im } f = \{y : y \in Y, \quad y = f(x) \quad \text{for some } x \in X\}. \tag{1.18}$$

The sine function has domain $\mathbb{R}$, maps $\mathbb{R}$ into $\mathbb{R}$ and has image equal to $\{y : -1 \leqslant y \leqslant 1\}$. As a further example if $X = \{p, q, r, s, t\}$ and if $Y = \{-1, 2, 4\}$ then a mapping $f : X \rightarrow Y$ is defined in letting $f(p) = -1$, $f(q) = -1$, $f(r) = 2$, $f(s) = 2$, $f(t) = 2$, here f has domain X and Im $f = \{-1, 2\}$.

Problem 1.15 Let $A = \{a, b, c\}$, $N = \{1, 2, 3, 4\}$. Let $g : A \rightarrow N$ defined by $g(a) = 2$, $g(b) = 3$, $g(c) = 3$. What is the domain and range of the mapping g?

Solution. By definition of g the domain is A. Clearly Im $g = \{2, 3\}$. $\square$

Notice that, although $f(x)$ is uniquely determined by x, many elements of X may be mapped into the same element of Y. For this reason f is sometimes called a *many–one* mapping.

Certain mappings play important roles in other branches of mathematics as well as in group theory. The examples used below to illustrate some of the concepts have not necessarily been taken from a group-theoretic or even from an algebraic context.

Problem 1.16 Let $X = \{x : x \in \mathbb{R}, \ 0 \leqslant x\}$ and let $f : X \rightarrow \mathbb{R}$ be defined by $f(x) = 2x/(x+1)$ $(x \in X)$. Prove that Im $f = \{y : y \in \mathbb{R}, 0 \leqslant y < 2\}$ and that if $x_1, x_2 \in X$ and $x_1 \neq x_2$ then $f(x_1) \neq f(x_2)$.

Solution. We prove first that Im $f \subseteq \{y : 0 \leqslant y < 2\}$. Let $x \in X$, then

7

$0 \leqslant x$ and hence

$$0 \leqslant \frac{2x}{x+1} = 2\left(1 - \frac{1}{x+1}\right) < 2. \tag{1.19}$$

Thus $f(x) \in \{y : 0 \leqslant y < 2\}$ and so $\operatorname{Im} f \subseteq \{y : 0 \leqslant y < 2\}$. To prove that $\operatorname{Im} f = \{y : 0 \leqslant y < 2\}$ we require to show that for any real number t, $0 \leqslant t < 2$, we can find a positive real number s such that $f(s) = t$. In other words does there exist a positive real number s such that $2s/(s+1) = t$ when $0 \leqslant t < 2$? By simple algebra $s = t/(2-t)$ and hence $s \geqslant 0$, thus we have $f(t/(2-t)) = t$ where $t/(2-t) \geqslant 0$. This establishes that $\operatorname{Im} f = \{y : 0 \leqslant y < 2\}$.

Let $x_1, x_2 \in X$ be such that $f(x_1) = f(x_2)$. We want to show that $x_1 = x_2$. We have $2x_1/(x_1+1) = 2x_2/(x_2+1)$ and, again by simple algebra, $x_1 = x_2$. This completes the solution. ☐

Let X, Y be nonempty sets and let $f : X \to Y$ be such that $x_1, x_2 \in X$, $x_1 \neq x_2$ implies that $f(x_1) \neq f(x_2)$, (equivalently $f(x_1) = f(x_2)$ implies $x_1 = x_2$), then f is called a *one–one mapping*. A one–one mapping is also called an *injection* or is said to be *injective*.

Problem 1.17 Let $X = \{s, a, g, e\}$ and let $Y = \{t, h, a, l, e, s\}$. Construct an injection f from X to Y.

Solution. Provided we define f so that $f(s)$, $f(a)$, $f(g)$, $f(e)$ are all different we have an injection. As a particular example let $f(s) = h$, $f(a) = a$, $f(g) = t$, $f(e) = s$, then f is an injection. ☐

Problem 1.18 Let $Y = \{y : y \in \mathbb{R}, \ 0 \leqslant y\}$ and let $f : \mathbb{R} \to Y$ be defined by $f(x) = x^2 \ (x \in \mathbb{R})$. Prove that $\operatorname{Im} f = Y$ but f is not a one–one mapping.

Solution. Let $y \in \mathbb{R}$, $y \geqslant 0$. Then $\sqrt{y}$ is a positive real number and so $f(\sqrt{y}) = (\sqrt{y})^2 = y$, hence $\operatorname{Im} f = Y$. f is not one–one since $f(-1) = (-1)^2 = 1 = f(1)$. ☐

Let X, Y be nonempty sets and let $f : X \to Y$ be such that $\operatorname{Im} f = Y$, then f is said to map X onto Y and f is called an *onto mapping*. An onto mapping is also called a *surjection* or is said to be *surjective*.

Thus if $f : X \to Y$ is a mapping f is injective if distinct elements of X have distinct images in Y under f and f is surjective if every element of Y is the image of some element of X.

Problem 1.19 Let $X = \{e, u, c, l, i, d\}$ and let $Y = \{w, i, s, e\}$. Construct a surjection f from X to Y.

Solution. Provided we define f so that w, i, s, e appear amongst $f(e)$, $f(u)$, $f(c)$, $f(l)$, $f(i)$, $f(d)$ we have a surjection. As a particular example let

$f(e) = w$, $f(u) = i$, $f(c) = s$, $f(l) = e$, $f(e) = s$, $f(d) = e$, then f is a surjection. $\qquad\qquad\qquad\qquad\qquad\qquad\qquad\qquad\qquad\qquad\qquad\quad\square$

A mapping that is both an injection and a surjection is called a *bijection* and is said to be *bijective*; in other words a bijective mapping is both one–one and onto.

Let X be a nonempty set, the mapping ι_X on X defined by $\iota_X(x) = x(x \in X)$ is called the *identity mapping* on X. The identity mapping on X leaves the elements of X fixed, ι_X is clearly bijective.

Notice that two mappings $f : X \to Y$ and $g : X \to Y$ are regarded as *equal* if and only if $f(x) = g(x)$ for all $x \in X$.

Problem 1.20 Construct a bijection f such that $f : \mathbb{R} \to \mathbb{R}$ but f is not the identity mapping on $\mathbb{R}$.

Solution. Examples of such bijections are easily found. For example, let $f(x) = x + 1$ ($x \in \mathbb{R}$), f is one–one and onto but $f \neq \iota_{\mathbb{R}}$ $\qquad\quad\square$

In trigonometry we become familiar with expressions like 'sin^2x', 'cos x', etc. We think of forming 'sin x' and then squaring the result. We put this notion into a functional or a mapping notation as follows: define mappings such that $f : Y \to Z$, $g : X \to Y$. Define a mapping h such that $f : \mathbb{R} \to \mathbb{R}$ and $g : \mathbb{R} \to \mathbb{R}$. Then we have $\sin^2 x = (\sin x)^2 = f(\sin x) = f(g(x))$, and we define a new mapping $h : \mathbb{R} \to \mathbb{R}$ by $h(x) = f(g(x)) = \sin^2 x$ ($x \in \mathbb{R}$).

More generally let X, Y, Z be three nonempty sets and let f, g be mappings such that $f : Y \to Z$, $g : X \to Y$. Define a mapping h such that $h : X \to Z$ by

$$h(x) = f(g(x)) \quad (x \in X). \tag{1.20}$$

We call h the *composite mapping* of g followed by f. The composite mapping is often denoted by the so-called *circle-notation* $\circ$, in which $h = f \circ g$ is that mapping of X into Z defined by

$$(f \circ g)(x) = f(g(x)) \quad (x \in X). \tag{1.21}$$

As an aid to memory it is useful to write this composition of mappings as

$$X \overset{g}{\to} Y \overset{f}{\to} Z \tag{1.22}$$

where the notation signifies that we first apply g to X and then f to Y.

Problem 1.21 Let f, g, h be the mappings of $\mathbb{R}$ into $\mathbb{R}$ defined by $f(x) = x + 1$ ($x \in \mathbb{R}$), $g(x) = x - 1$ ($x \in \mathbb{R}$), $h(x) = 3x$ ($x \in \mathbb{R}$). Evaluate $f \circ g$, $g \circ f$, $f \circ h$, $h \circ f$.

Solution. Let $x \in \mathbb{R}$ then, by equation 1.21,

$$(f \circ g)(x) = f(g(x)) = f(x-1) = (x-1)+1 = x$$
$$(g \circ f)(x) = g(f(x)) = g(x+1) = (x+1)-1 = x$$
$$(f \circ h)(x) = f(h(x)) = f(3x) = 3x+1$$
$$(h \circ f)(x) = h(x+1) = 3(x+1) = 3x+3. \qquad \square$$

Problem 1.22 Let X, Y, Z be nonempty sets and let f, g be mappings such that $f : Y \to Z$, $g : X \to Y$. Prove (i) that if f, g are one–one then so is $f \circ g$ and (ii) that if f, g are onto then so is $f \circ g$.

Solution. We have $X \xrightarrow{g} Y \xrightarrow{f} Z$.

(i) Suppose f, g are one–one. Let x_1, $x_2 \in X$ be such that $(f \circ g)(x_1) = (f \circ g)(x_2)$. Then $f(g(x_1)) = f(g(x_2))$ and, since f is one–one we have $g(x_1) = g(x_2)$. Since g is one–one $x_1 = x_2$. Thus $f \circ g$ is one–one.

(ii) Suppose f, g are onto. Let $z \in Z$. Then, since f is onto, there exists $y \in Y$ such that $f(y) = z$. Since g is onto there exists $x \in X$ such that $g(x) = y$. Thus $(f \circ g)(x) = f(g(x)) = f(y) = z$ and so $f \circ g$ is onto. $\qquad \square$

Problem 1.23 Let $W = \{f, a, i, t, h\}$, $X = \{h, o, p, e\}$, $Y = \{c, h, a, r, i, t, y\}$ and $Z = \{l, o, v, e\}$. Mappings g_1, g_2, g_3 with domains Y, X, W respectively are defined by

$$g_3(f) = h, \quad g_3(a) = o, \quad g_3(i) = p, \quad g_3(t) = g_3(h) = e;$$
$$g_2(h) = c, \quad g_2(o) = r, \quad g_2(p) = t, \quad g_2(e) = y;$$
$$g_1(c) = g_1(h) = l, \quad g_1(a) = g_1(i) = g_1(y) = v,$$
$$g_1(r) = g_1(t) = e.$$

Evaluate whichever of the following exist $g_1 \circ g_2$, $g_2 \circ g_3$, $g_1 \circ g_3$, $g_2 \circ g_2$. Verify that $(g_1 \circ g_2) \circ g_3$ and $g_1 \circ (g_2 \circ g_3)$ exist and are equal.

Solution. We observe that we have

$$W \xrightarrow{g_3} X \xrightarrow{g_2} Y \xrightarrow{g_1} Z \qquad (1.23)$$

and therefore we can 'read off' the existing mappings, i.e. all except $g_1 \circ g_3$ and $g_2 \circ g_2$. We have

$$(g_1 \circ g_2)(h) = g_1(c) = l \qquad (g_2 \circ g_3)(f) = g_2(h) = c$$
$$(g_1 \circ g_2)(o) = g_1(r) = e \qquad (g_2 \circ g_3)(a) = g_2(o) = r$$
$$(g_1 \circ g_2)(p) = g_1(t) = e \qquad (g_2 \circ g_3)(i) = g_2(p) = t$$
$$(g_1 \circ g_2)(e) = g_1(y) = v \qquad (g_2 \circ g_3)(t) = g_2(e) = y$$
$$(g_2 \circ g_3)(h) = g_2(e) = y.$$

Applying equation 1.21 again we have

$$[(g_1 \circ g_2) \circ g_3](f) = (g_1 \circ g_2)(g_3(f)) = (g_1 \circ g_2)(h) = 1$$
$$[g_1 \circ (g_2 \circ g_3)](f) = g_1((g_2 \circ g_3)(f)) = g_1(c) = 1$$

and so

$$[(g_1 \circ g_2) \circ g_3](f) = g_1 \circ (g_2 \circ g_3)(f).$$

Repeating this argument shows that $(g_1 \circ g_2) \circ g_3 = g_1 \circ (g_2 \circ g_3)$. ☐

Problem 1.24 Let W, X, Y, Z be four nonempty sets and let g_1, g_2, g_3 be mappings as follows:

$$W \xrightarrow{g_3} X \xrightarrow{g_2} Y \xrightarrow{g_1} Z. \tag{1.24}$$

Prove that

$$(g_1 \circ g_2) \circ g_3 = g_1 \circ (g_2 \circ g_3). \tag{1.25}$$

Solution. Let $w \in W$. We apply equation 1.21 several times, thus

$$[(g_1 \circ g_2) \circ g_3](w) = [g_1 \circ g_2](g_3(w)) = g_1[g_2(g_3(w))]$$
$$= g_1\{(g_2 \circ g_3)(w)\} = \{g_1 \circ (g_2 \circ g_3)\}(w).$$

Since w is arbitrary we obtain the result. ☐

It is easy to show that if g_1, g_2, g_3 are three mappings such that either $(g_1 \circ g_2) \circ g_3$ exists or $g_1 \circ (g_2 \circ g_3)$ exists then both exist and are consequently equal. We express the equality in equation 1.25 by saying that the circle-composition of mappings is *associative*.

Problem 1.25 Let $X = \{w, i, s, d, o, m\}$ and $Y = \{s, a, n, i, t, y\}$. Let $f : X \to Y$ and $g : Y \to X$ be defined by $f(w) = y$, $f(i) = a$, $f(s) = n$, $f(d) = i$, $f(o) = t$, $f(m) = s$; $g(s) = m$, $g(a) = i$, $g(n) = s$, $g(i) = d$, $g(t) = o$, $g(y) = w$. Prove that $g \circ f = \iota_X$ and $f \circ g = \iota_X$.

Solution. The verification that $g \circ f = \iota_X$ and that $f \circ g = \iota_X$ should now be routine; for example,

$$(g \circ f)(w) = g(f(w)) = g(y) = w = \iota_X(w). \qquad ☐$$

Let X, Y be nonempty sets and let f, g be mappings such that $f : X \to Y$ and $g : Y \to X$. If $g \circ f = \iota_X$ and $f \circ g = \iota_Y$ then we call f and g *inverse mappings*.

Problem 1.26 Let $f : X \to Y$ and $g : Y \to X$ be inverse mappings. Prove that f, g are bijections.

Solution. We prove only that f is a bijection, the proof that g is a bijection being mere repetition. Let $x_1, x_2 \in X$ and suppose $f(x_1) = f(x_2)$.

Then

$$x_1 = \iota_X(x_1) = (g \circ f)(x_1) = g(f(x_1)) = g(f(x_2))$$
$$= (g \circ f)(x_2) = \iota_X(x_2) = x_2. \tag{1.26}$$

and so f is one–one. Let $y \in Y$, we have to show that there exists $x \in X$ such that $f(x) = y$. Now $g(y) \in X$ and thus $f(g(y)) = (f \circ g)(y) = \iota_Y(y) = y$. Hence if $x = g(y)$ we have $f(x) = y$ and so f is onto. ☐

Problem 1.27 Let X, Y be nonempty sets and let $f : X \to Y$ be a bijection. Prove that there exists $g : Y \to X$ such that f, g are inverse mappings.

Solution. Let $y \in Y$. Since f is surjective there exists $x \in X$ such that $f(x) = y$ and since f is injective x is uniquely determined by y. We define $g : Y \to X$ by letting $g(y) = x$. Then g is a properly defined mapping and furthermore $(g \circ f)(x) = g(f(x)) = g(y) = x$, $(f \circ g)(y) = f(g(y)) = f(x) = y$. Thus f, g are inverse mappings. ☐

Problem 1.28 Let A, B, C be nonempty sets and let $f : A \to B$ and $g : B \to C$ be mappings. Prove that if $g \circ f$ is an injection then f is an injection and if $g \circ f$ is a surjection then g is a surjection.

Solution. We have

$$A \xrightarrow{f} B \xrightarrow{g} C. \tag{1.27}$$

Suppose $g \circ f$ is an injection. Let $a_1, a_2 \in A$ and suppose $f(a_1) = f(a_2)$, we have to show that $a_1 = a_2$. We have $(g \circ f)(a_1) = g(f(a_1)) = g(f(a_2)) = (g \circ f)(a_2)$ and, as $g \circ f$ is an injection, it follows that $a_1 = a_2$. Suppose $g \circ f$ is a surjection. Let $c \in C$, we have to show that there exists $b \in B$ such that $g(b) = c$. Since $g \circ f$ is a surjection there exists $a \in A$ such that $(g \circ f)(a) = c$. Thus we let $b = f(a)$ for then $g(b) = g(f(a)) = (g \circ f)(a) = c$.

☐

1.3 Equivalence relations Given any two real numbers r_1 and r_2 we have the relation of 'less than or equal to' between them since either $r_1 \leqslant r_2$ or $r_2 \leqslant r_1$. We may envisage other relations on $\mathbb{R}$, thus we might say that if $s_1, s_1 \in \mathbb{R}$ then s_1 is related to s_2 if and only if $s_1 = ns_2$ for some integer n depending on s_1 and s_2. For notational convenience we introduce a symbol $\sim$ meaning 'stands in the given relation to', thus $s_1 \sim s_2$ if and only if there exists $n \in \mathbb{Z}$ such that $s_1 = ns_2$, we have $3 \sim 1\frac{1}{2}$ since $3 = 2 \times 1\frac{1}{2}$ but we do not have $\frac{1}{2} \sim 3$ since $\frac{1}{2}$ is not an integral multiple of 3.

In more precise language a relation, denoted by $\sim$, on a set S is obtained from any subset T of $S \times S$ by letting $s_1 \sim s_2$ if and only if $(s_1, s_2) \in T$.

Problem 1.29 A relation $\sim$ is defined on the complex numbers $\mathbb{C}$ by letting $z_1 \sim z_2$ $(z_1, z_2 \in \mathbb{C})$ if and only if $|z_1| = |z_2|$. Prove that $z \sim z$ $(z \in \mathbb{C})$ and that $z_1 \sim z_2$ $(z_1, z_2 \in \mathbb{C})$ implies that $z_2 \sim z_1$ $(z_1, z_2 \in \mathbb{C})$.

Solution. The desired conclusions are immediate since $z \sim z$ by definition and $|z_1| = |z_2|$ implies $|z_2| = |z_1|$. $\qquad\square$

A relation $\sim$ on a set S is called *reflexive* if $s \sim s$ for all $s \in S$ and is called *symmetric* if $s_1 \sim s_2$ $(s_1, s_2 \in S)$ implies $s_2 \sim s_1$. The relation of inequality on $\mathbb{R}$ is reflexive since $a \leqslant a$ $(a \in \mathbb{R})$ but is not symmetric since $b \leqslant c$ $(b, c \in \mathbb{R})$ does not imply $c \leqslant b$.

Problem 1.30 A relation $\sim$ is defined on the rationals $\mathbb{Q}$ by letting $a \sim b$ $(a, b \in \mathbb{Q})$ if and only if $a - b \geqslant 1$. Show that $\sim$ is neither reflexive nor symmetric. Prove that if $u, v, w \in \mathbb{Q}$ are such that $u \sim v$ and $v \sim w$ then $u \sim w$.

Solution. It is false that $a - a \geqslant 1$ and so we do not have $a \sim a$ $(a \in \mathbb{Q})$. If $a - b \geqslant 1$ then it is false that $b - a \geqslant 1$ and so $a \sim b$ $(a, b \in \mathbb{Q})$ does not imply that $b \sim a$. If $u \sim v$ and $v \sim w$ $(u, v, w \in \mathbb{Q})$ we have $u - w = (u-v) + (v-w) \geqslant 1 + 1 \geqslant 1$ and so $u \sim w$. $\qquad\square$

A relation $\sim$ on a set S is called *transitive* if $s_1 \sim s_2$ and $s_2 \sim s_3$ $(s_1, s_2, s_3 \in S)$ implies $s_1 \sim s_3$. The relation of inequality on $\mathbb{R}$ is transitive since $a \leqslant b$ and $b \leqslant c$ $(a, b, c \in \mathbb{R})$ implies $a \leqslant c$.

A relation $\sim$ on a set S that is reflexive, symmetric and transitive is called an *equivalence relation*.

Problem 1.31 Prove that the relation $\sim$ defined on $\mathbb{Q}$ as follows is an equivalence relation. For $a, b \in \mathbb{Q}$ let $a \sim b$ if and only if there exist m, $n \in \mathbb{Z}$, $m \neq 0$, $n \neq 0$, such that $a^m = b^n$.

Solution. The relation is reflexive for $a^1 = a^1$ implies $a \sim a$ $(a \in \mathbb{Q})$. The relation is symmetric by definition. Let now $a \sim b$ and $b \sim c$ $(a, b \in \mathbb{Q})$. Then there exist $m, n, r, s \in \mathbb{Z}$, $m \neq 0$, $n \neq 0$, $r \neq 0$, $s \neq 0$ such that $a^m = b^n$ and $b^r = c^s$. Then

$$a^{mr} = (a^m)^r = (b^n)^r = b^{rn} = (b^r)^n = (c^s)^n = c^{sn} \qquad (1.28)$$

and so $\sim$ is transitive. Hence, finally, $\sim$ is an equivalence relation. $\qquad\square$

Problem 1.32 Let S be a set on which an equivalence relation $\sim$ is defined. For $s \in S$ define $\mathscr{C}(s) = \{x : x \in S, \; s \sim x\}$. Prove that $s \in \mathscr{C}(s)$ and that if $t \in S$ then either $\mathscr{C}(s) = \mathscr{C}(t)$ or $\mathscr{C}(s) \cap \mathscr{C}(t)$ is empty.

Solution. Since $s \sim s$ we have $s \in \mathscr{C}(s)$. For the sake of argument suppose $\mathscr{C}(s) \cap \mathscr{C}(t) \neq \varnothing$, we shall prove $\mathscr{C}(s) = \mathscr{C}(t)$. Let $y \in \mathscr{C}(s) \cap \mathscr{C}(t)$, then $s \sim y$ and $t \sim y$. Since $\sim$ is reflexive $y \sim s$ and so from $t \sim y$ and $y \sim s$ we deduce, by transitivity, that $t \sim s$. Let $w \in \mathscr{C}(s)$, then $s \sim w$. As $t \sim s$ and $s \sim w$ we have $t \sim w$ and so $w \in \mathscr{C}(t)$. Thus $\mathscr{C}(s) \subseteq \mathscr{C}(t)$ and similarly we prove $\mathscr{C}(t) \subseteq \mathscr{C}(s)$. The desired conclusion is then immediate. $\qquad\square$

Let S be a set on which an equivalence relation $\sim$ is defined. The set $\mathscr{C}(s) = \{x : x \in S, \; s \sim x\}$ is called the *equivalence class* containing s. An equivalence class is determined by any element belonging to it. S is the union of its equivalence classes and, since any two equivalence classes either coincide or have empty intersection, S is said to be the *disjoint union* of its equivalence classes. Of course these equivalence classes depend on the particular equivalence relation, different relations would normally yield different equivalence classes.

Problem 1.33 Let $S = \{1, 2, 3, 4, 5, 6, 7, 8, 9, 10\}$. A relation $\sim$ is defined on S by $a \sim b \ (a, b \in S)$ if and only if $a - b$ is divisible by 4. Prove that $\sim$ is an equivalence relation and determine the equivalence classes.

Solution. Clearly $a \sim a \ (a \in S)$ for $0 = a - a$ is trivially divisible by 4. If $a - b$ is divisible by 4 $(a, b \in S)$ then so is $b - a$ and hence $a \sim b$ implies $b \sim a$. If $a \sim b$ and $b \sim c \ (a, b, c \in S)$ then $a - b$, $b - c$ are divisible by 4 and so $a - c$ is divisible by 4 since $a - c = (a - b) + (b - c)$.

The equivalence classes are $\{1, 5, 9\}$, $\{2, 6, 10\}$, $\{3, 7\}$, $\{4, 8\}$. $\qquad\square$

EXERCISES

1. Let $A = \{1, 2. 3, 6, 7\}, B = \{3, 4, 5, 6\}, C = \{1, 2, 3, 4, 6\}, D = \{3, 5, 6, 7\}$. Prove that $A \cup B = C \cup D$ and $A \cap B = C \cap D$.

2. For any real number t let $A_t = \{x : t \leqslant x\}$. Prove that $\bigcup_{t \in \mathbb{R}} A_t = \mathbb{R}$ and that $\bigcap_{t \in \mathbb{R}} A_t = \varnothing$.

3. Let A, B, C be sets. Prove that $A \cup (B \cap C) = (A \cup B) \cap (A \cup C)$.

4. [Hard] Let X, Y, S, T be sets such that $X \cup Y = T \cup S$ and $X \cap Y = T \cap S = \varnothing$. Prove that $X = \varnothing$ if and only if $T = (X \cap S) \cup (Y \cap T)$.

5. Let $f : \mathbb{R} \to \mathbb{R}$ be defined by $f(x) = x^2/(x^2 + 1) \ (x \in \mathbb{R})$. Prove that $f(\mathbb{R}) = \{y : 0 \leqslant y < 1\}$ and that f is not injective.

6. Let X, Y be finite sets having m, n distinct elements respectively. Prove that there are n^m distinct mappings from X to Y. If one of these mappings is a bijection prove that $m = n$ and that there are exactly $n!$ bijections.

7. Let A, B, C be nonempty sets and let $f : A \to B$ and $g : B \to C$ be

mappings. Construct an example in which f is onto, g is one–one and $g \circ f$ is neither onto nor one–one.

8.   A relation $\sim$ is defined in $\mathbb{Z}$ by letting $a \sim b$ $(a, b \in \mathbb{Z})$ if and only if 6 divides $a - b$. Prove that $\sim$ is an equivalence relation.

9.   A relation $\sim$ is defined in $\mathbb{C}$ by letting $z_1 \sim z_2$ $(z_1, z_2 \in \mathbb{C})$ if and only if $z_1 - z_2 \in \mathbb{R}$. Prove that $\sim$ is an equivalence relation.

Chapter 2

Semigroups, Groups

2.1 Semigroups Let S be a nonempty set. Suppose we have a rule $*$ by which, for every ordered pair (s_1, s_2) of elements of S, we can determine a third element of S, denoted by $s_1 * s_2$. Then we say that S is *closed* under a *law of composition* $*$ on S. Thus on the real numbers $\mathbb{R}$ we have (at least) two laws of composition, namely multiplication $\times$ and addition $+$, for it is trivial that if $s_1, s_2 \in \mathbb{R}$ then $s_1 \times s_2 \in \mathbb{R}$ and $s_1 + s_2 \in \mathbb{R}$.

Let S be closed under a law of composition $*$. Then for every ordered triple (s_1, s_2, s_3) of elements of S we may assign two possible products. We may determine first $s_1 * s_2$ and then $(s_1 * s_2) * s_3$ or, on the other hand, we may determine first $s_2 * s_3$ and then $s_1 * (s_2 * s_3)$. If for all choices $s_1, s_2, s_3 \in S$ we have

$$(s_1 * s_2) * s_3 = s_1 * (s_2 * s_3) \tag{2.1}$$

then $*$ is said to be an *associative* law of composition and S is called a *semigroup* under the law of composition $*$.

Problem 2.1 Prove that $\mathbb{R}$ is a semigroup under either multiplication or addition as the law of composition.

Solution. Since $\mathbb{R}$ is closed under both laws we have merely to observe that for any real numbers s_1, s_2, s_3

$$(s_1 \times s_2) \times s_3 = s_1 \times (s_2 \times s_3) \tag{2.2}$$
$$(s_1 + s_2) + s_3 = s_1 + (s_2 + s_3). \qquad \square \tag{2.3}$$

Problem 2.2 Let X be a nonempty set and let S be the set of mappings of X into X. Prove that S is a semigroup under the circle composition of mappings.

Solution. Let $f, g \in S$, then $f : X \to X$ and $g : X \to X$. By the definition (1.21) of the circle-composition of f and g it follows that $f \circ g \in S$. By Problem 1.24 with $W = X = Y = Z$ it follows from equation 1.25 that the circle-composition of mappings is associative. $\square$

Problem 2.3 Let S be a nonempty set and let $a \in S$. Prove that S is a semigroup if we define $s * t = a$ $(s, t \in S)$.

Solution. S is closed under $*$ and if $s_1, s_2, s_3 \in S$ we have

$$(s_1 * s_2) * s_3 = a * s_3 = a = s_1 * a = s_1 * (s_2 * s_3).$$

Thus S is a semigroup under $*$.

This problem shows how any nonempty set may be regarded, in a trivial way, as a semigroup. □

Although this book is concerned with associative laws of composition it should not be imagined that nonassociative laws are either not useful or do not exist outside a purely mathematical context.

Problem 2.4 Let V be the set of three-dimensional geometric vectors and let the law of composition on V be the usual vector product. Prove that this law is nonassociative by showing that if $\mathbf{u}, \mathbf{v}, \mathbf{w} \in V$ then

$$(\mathbf{u} \wedge \mathbf{v}) \wedge \mathbf{w} = \mathbf{u} \wedge (\mathbf{v} \wedge \mathbf{w})$$

if and only if $\mathbf{v} \wedge (\mathbf{w} \wedge \mathbf{u}) = 0$.

Solution. Expanding the triple vector products we have

$$\mathbf{u} \wedge (\mathbf{v} \wedge \mathbf{w}) = (\mathbf{u} . \mathbf{w})\mathbf{v} - (\mathbf{u} . \mathbf{v})\mathbf{w} \qquad (2.4)$$

$$(\mathbf{u} \wedge \mathbf{v}) \wedge \mathbf{w} = (\mathbf{w} . \mathbf{u})\mathbf{v} - (\mathbf{w} . \mathbf{v})\mathbf{u}. \qquad (2.5)$$

Thus the left-hand sides of equations 2.4 and 2.5 are equal if and only if

$$\mathbf{v} \wedge (\mathbf{w} \wedge \mathbf{u}) = (\mathbf{v} . \mathbf{u})\mathbf{w} - (\mathbf{v} . \mathbf{w})\mathbf{u} = 0$$

which is evidently not true for arbitrary $\mathbf{u}, \mathbf{v}, \mathbf{w}$. □

If the semigroup S has a finite number of elements, $S = \{s_1, s_2, \ldots, s_n\}$, say, we can express the formation of the n^2 elements $s_i * s_j$ in the form of an array:

	s_1	s_2	$\cdots$	s_j	$\cdots$	s_n
s_1	$s_1 * s_1$	$s_1 * s_2$	$\cdots$	$s_1 * s_j$		$s_1 * s_n$
s_2	$s_2 * s_1$	$s_2 * s_2$	$\cdots$	$s_2 * s_j$		$s_2 * s_n$
⋮		·	·	·		
s_i	$s_i * s_1$	$s_i * s_2$	$\cdots$	$s_i * s_j$		$s_i * s_n$
⋮		·	·	·		
s_n	$s_n * s_1$	$s_i * s_2$	$\cdots$	$s_n * s_j$		$s_n * s_n$

(2.6)

where the entry in the ith row and the jth column is the element $s_i * s_j$.

Problem 2.5 If $i = \sqrt{-1}$, write down the array for the four-element semigroup $S = \{1, i, -1, -i\}$ under the usual multiplication.

17

Solution. The array is:

	1	i	-1	$-i$
1	1	i	-1	$-i$
i	i	-1	$-i$	1
-1	-1	$-i$	1	i
$-i$	$-i$	1	i	-1

(2.7)

$\square$

In order to simplify notation we frequently omit the symbol $*$ and we write $s_1 s_2$, the law of composition being understood; of course if a second law of composition arises then a symbol must be inserted (in the set of real numbers $\mathbb{R}$ we customarily omit the multiplication sign but retain the sign for addition). By analogy with the real numbers we call $s_1 s_2$ the *product* of s_1 and s_2 and the law of composition *multiplication*. Thus multiplication is associative on S if, for $s_1, s_2, s_3 \in S$,

$$(s_1 s_2)s_3 = s_1 (s_2 s_3).$$ (2.8)

If the symbol $+$ is used we call $s_1 + s_2$ the *sum* of s_1 and s_2 and the law of composition *addition*. Thus addition is associative on S if, for $s_1, s_2, s_3 \in S$

$$(s_1 + s_2) + s_3 = s_1 + (s_2 + s_3).$$ (2.9)

Problem 2.6 Let S be the four-element set $\{a, b, c, d\}$ with multiplication:

	a	b	c	d
a	a	a	c	d
b	b	b	c	d
c	c	c	c	c
d	d	d	d	d

(2.10)

Determine the following sets $\{x \in S : x^2 = x\}$, $\{x \in S : xa = ax\}$, $\{x \in S : xd = a\}$ and evaluate the products $(ab)(cd)$, $((ab)c)d$, $(a(bc))d$.

Solution.

$$\{x \in S : x^2 = x\} = \{a, b, c, d\} = S$$
$$\{x \in S : xa = ax\} = \{a, c, d\}$$
$$\{x \in S : xd = a\} = \varnothing$$
$$(ab)(cd) = ac = c$$
$$((ab)c)d = (ac)d = cd = c$$
$$(a(bc))d = (ac)d = cd = c.$$

$\square$

It may be verified that S in Problem 2.6 is a semigroup and then the last three equalities are, in fact, particular examples of a general result, namely

18

that in any semigroup S the product of any n elements taken in a given order is independent of how we evaluate the product. Thus, for example, considering possible ordered products of $s_1, s_2, s_3, s_4, s_5 \in S$ we have

$$\begin{aligned}
((s_1 s_2)s_3)(s_4 s_5) &= (s_1(s_2(s_3 s_4)))s_5 \\
&= (((s_1 s_2)s_3)s_4)s_5.
\end{aligned} \tag{2.11}$$

It follows that the bracketing may be omitted and, in the above instance, we would write $s_1 s_2 s_3 s_4 s_5$. In the general case we write $s_1 s_2 \ldots s_n$ where the meaning is now unambiguous. We extend this notion to the product $ss \ldots s$ of $s \in S$ taken n times with itself and we write

$$s^n = ss \ldots s. \tag{2.12}$$

The index law is then an easy consequence, namely

$$s^m s^n = s^{m+n} \quad (m, n = 1, 2, \ldots). \tag{2.13}$$

Problem 2.7 Let S be a semigroup. Let $e, f \in S$ be such that $es = s$, $tf = t$ for all $s, t \in S$. Prove that $e = f$.

Solution. We observe that, in particular, we may replace s by f and t by e in the above to obtain $ef = f$ and $ef = e$. This yields the result. □

A semigroup S possessing an element e such that $es = se = s$ for all $s \in S$ is called a *monoid*. By the above the element e (if it exists) is unique and we call e the *identity* of S.

Problem 2.8 Let S be a monoid with identity e. Let $s \in S$ and *suppose* there exists $s', s'' \in S$ such that $s's = ss'' = e$. Prove $s' = s''$.

Solution.

$$s' = s'e = s'(ss'') = (s's)s'' = es'' = s''.$$

(Notice the crucial reliance on associativity.) □

If S is a monoid with identity e, the unique element $s' \in S$ such that $s's = ss' = e$ is called the *inverse* of s (s' may not exist).

Problem 2.9 Let S be the multiplicative semigroup of nonzero integers. Which elements of S have inverses?

Solution. The identity of S is $+1$ and only $+1$ and -1 have inverses. □

Problem 2.10 Let S be the multiplicative semigroup of nonzero real numbers. Prove that every element of S has an inverse.

Solution. The identity is $+1$ and if s is a nonzero number so is $1/s$ and

$$\left(\frac{1}{s}\right)s = s\left(\frac{1}{s}\right) = 1.$$

Thus s has $1/s$ as inverse. □

19

A monoid S in which an inverse exists for every element of S is called a *group*; if S is multiplicative we write x^{-1} for the inverse of $x \in S$.

Two elements s, t of a semigroup S such that $st = ts$ are said to *commute*; if *any* two elements of S commute then S is called *commutative*. For example the set of integers $\mathbb{Z}$ with addition as the law of composition forms a commutative group.

Problem 2.11 Prove that the semigroup S in Problem 2.6 is not commutative.

Solution. The semigroup is not commutative since
$$ab = a \neq b = ba.$$
□

Problem 2.12 Prove that the set S of matrices of the form
$$\begin{pmatrix} \cos\theta & \sin\theta \\ -\sin\theta & \cos\theta \end{pmatrix} \quad (\theta \in \mathbb{R})$$
under the usual matrix multiplication, is a commutative group.

Solution. We must first verify that S is closed under the multiplication. This follows from elementary trigonometry since
$$\begin{pmatrix} \cos\alpha & \sin\alpha \\ -\sin\alpha & \cos\alpha \end{pmatrix}\begin{pmatrix} \cos\beta & \sin\beta \\ -\sin\beta & \cos\beta \end{pmatrix}$$
$$= \begin{pmatrix} \cos\alpha\cos\beta - \sin\alpha\sin\beta & \cos\alpha\sin\beta + \sin\alpha\cos\beta \\ -\sin\alpha\cos\beta - \cos\alpha\sin\beta & -\sin\alpha\sin\beta + \cos\alpha\cos\beta \end{pmatrix}$$
$$= \begin{pmatrix} \cos(\alpha+\beta) & \sin(\alpha+\beta) \\ -\sin(\alpha+\beta) & \cos(\alpha+\beta) \end{pmatrix} \quad (\alpha, \beta \in \mathbb{R}) \tag{2.14}$$
which is of the required form. Matrix multiplication is associative and
$$\begin{pmatrix} 1 & 0 \\ 0 & 1 \end{pmatrix} = \begin{pmatrix} \cos 0 & \sin 0 \\ -\sin 0 & \cos 0 \end{pmatrix}$$
is the identity of S. Letting $\beta = -\alpha$ in equation 2.14 we have
$$\begin{pmatrix} \cos\alpha & \sin\alpha \\ -\sin\alpha & \cos\alpha \end{pmatrix}\begin{pmatrix} \cos(-\alpha) & \sin(-\alpha) \\ -\sin(-\alpha) & \cos(-\alpha) \end{pmatrix} = \begin{pmatrix} \cos 0 & \sin 0 \\ -\sin 0 & \cos 0 \end{pmatrix} = \begin{pmatrix} 1 & 0 \\ 0 & 1 \end{pmatrix}$$
$$= \begin{pmatrix} \cos(-\alpha) & \sin(-\alpha) \\ -\sin(-\alpha) & \cos(-\alpha) \end{pmatrix}\begin{pmatrix} \cos\alpha & \sin\alpha \\ -\sin\alpha & \cos\alpha \end{pmatrix} \tag{2.15}$$
from which we infer that all elements of S have inverses in S. Hence S is a group. On interchanging α and β in equation 2.14 we see that S is commutative.
□

Problem 2.13 The four-element sets $S_1 = \{e_1, a_1, b_1, c_1\}$ and

20

$S_2 = \{e_2, a_2, b_2, c_2\}$ have laws of composition $*$, $\circ$ respectively defined as follows:

$*$	e_1	a_1	b_1	c_1
e_1	e_1	a_1	b_1	c_1
a_1	a_1	b_1	c_1	e_1
b_1	b_1	c_1	e_1	a_1
c_1	c_1	e_1	a_1	b_1

$\circ$	e_2	a_1	b_2	c_2
e_2	e_2	a_2	b_2	c_2
a_2	a_2	b_2	c_2	e_2
b_2	b_2	a_2	e_2	c_2
c_2	c_2	e_2	a_1	b_2

(2.16)

Which of these sets is a group?

Solution. Let us first note that both multiplications are closed, that S_1 and S_2 appear to have identities e_1 and e_2 respectively (e.g. $e_1 * b_1 = b_1 * e_1 = b_1$) and that inverses appear to exist (e.g. $a_2 \circ c_2 = e_2 = c_2 \circ a_2$). Nevertheless S_2 is not a group for the multiplication is not associative as we have

$$b_2 \circ (a_2 \circ c_2) = b_2 \circ e_2 = b_2 \neq e_2 = a_2 \circ c_2 = (b_2 \circ a_2) \circ c_2.$$

By direct, if somewhat tedious, verification it may be shown that multiplication in S_1 is associative and so S_1 is a group. ☐

Problem 2.14 Let S be a semigroup. Prove that S is a group if both of the following conditions are satisfied:

   (i)   there exists $e \in S$ such that $es = s$ for all $s \in S$, and   (2.17)
   (ii) for all $t \in S$ there exists $t' \in S$ such that $t't = e$.   (2.18)

Solution. We show first that e is the identity of S. Let $x \in S$; then by (ii) there exists $x' \in S$ such that $x'x = e$ and, again by (ii), there exists $x'' \in S$ such that $x''x' = e$. By repeated applications of (i) and of associativity we have

$$xe = e(xe) = (x''x')(xe)$$
$$= x''(x'x)e = x''(e)e$$
$$= x''(ee) = x''e = x''(x'x) = (x''x')x$$
$$= ex = x$$

This result, together with (i), implies that e is the identity of S and so S is a monoid. To complete the proof that S is a group it is sufficient to show that $x'' = x$ and this is now immediate from Problem 2.8. ☐

Problem 2.14 yields a convenient minimal definition for a group. Notice that in the conditions as stated both e and t' are 'on the left'; a similar pair of conditions for a semigroup to be a group can be formulated 'on the right'.

Problem 2.15 Construct a semigroup S satisfying the following conditions:

(i)   there exists $e \in S$ such that $es = s$ for all $s \in S$, and   (2.19)

(ii) for all $t \in S$ there exists $t' \in S$ such that $tt' = e$, and   (2.20)

(iii) S is not a group.   (2.21)

Solution. We know that a group satisfies conditions 2.19 and 2.20 but, by condition 2.21, we are explicitly required to construct a semigroup that is not a group. Thus we are faced with the frequently occurring problem of existence, i.e. in this instance does such a semigroup exist? We try to construct as simple an example as possible. This leads us (among other possibilities) to consider the two-element semigroup $S = \{e, f\}$ having multiplication

	e	f
e	e	f
f	e	f

It is easy to verify that S is a semigroup satisfying conditions 2.19, 2.20 and 2.21.   ◻

Problem 2.16 Let G be a group and let $a, x, y \in G$ be such that $ax = ay$. Prove that $x = y$.

Solution. Since G is a group, G has an identity e and there exists $a^{-1} \in G$ such that $a^{-1}a = e$. As e is the identity and by applying equation 2.8 we have
$$x = ex = (a^{-1}a)x = a^{-1}(ax) = a^{-1}(ay) = (a^{-1}a)y = y.$$

In the same way we may deduce from $xa = ya$ that $x = y$. Thus we may 'cancel' equal elements from both sides of an equation involving the elements of a group.   ◻

Problem 2.17 Let G be a group and let $a, b \in G$. Prove that the equation $ax = b$ has the unique solution $x = a^{-1}b$.

Solution. We verify first that $a^{-1}b$ satisfies the equation. This follows from equation 2.8 since, if e is the identity of G, $a(a^{-1}b) = (aa^{-1})b = eb = b$. Furthermore if $ax = b(x \in G)$ then $ax = a(a^{-1}b)$ and so, by Problem 2.16, $x = a^{-1}b$.

Correspondingly the equation $ya = b$ has the unique solution $y = ba^{-1}$. Thus in a group we may 'divide' any element by any other; this possibility of 'division' is a characteristic feature of groups. The semigroup S consisting of the nonzero integers under multiplication does not always permit division, we cannot, for example solve $2x = 3$ for an $x \in S$; on the other hand we may also 'cancel' in S although this is not possible in all semigroups (cf. Problem 2.3).   ◻

Problem 2.18   Let S be a semigroup in which for any $a, b \in S$ there exists $x, y \in S$ such that $ax = b$ and $ya = b$. Prove that S is a group.

Solution. We apply the criterion afforded by Problem 2.14. Let $a \in S$. By hypothesis there exists $e \in S$ such that $ea = a$. We want to show that e satisfies condition 2.17 of Problem 2.14 but at this stage of our argument e might depend on a. Let $s \in S$. Then there exists $u \in S$ such that $au = s$ and so we have $es = e(au) = (ea)u = au = s$. Furthermore there exists $s' \in S$ such that $s's = e$. Hence, by Problem 2.14, G is a group.   □

2.2 Introduction to groups   We recall, from the previous section, that a nonempty set G closed under an associative law of composition is called a group if G has an (unique) identity e such that every element x of G has an (uniquely determined) inverse x^{-1} with respect to this identity.

By direct arguments it is sometimes possible to obtain the law of composition for a group G if it has a small number of elements.

Problem 2.19   Obtain the law of composition for a group G that has (i)1, (ii)2, (iii)3, (iv)4 elements.

Solution. (i) If G has only one element then this element must be the identity e and we have the trivial multiplication

$$
\begin{array}{c|c}
 & e \\
\hline
e & e
\end{array}
\tag{2.22}
$$

(ii) Let $G = \{e, a\}$. Since $a^2 \in G$ we must have $a^2 = e$ or $a^2 = a$ but $a^2 = a$ implies that $a = e$, thus $a^2 = e$ and we have

$$
\begin{array}{c|cc}
 & e & a \\
\hline
e & e & a \\
a & a & e
\end{array}
\tag{2.23}
$$

(iii) Let $G = \{e, a, b\}$. We consider ab and we observe that $ab \neq a$ for $ab = a$ implies $b = e$. Similarly $ab \neq b$ and hence $ab = e$. This implies that $b = a^{-1}$ and so $ba = e$. Hence we obtain the partially filled array

$$
\begin{array}{c|ccc}
 & e & a & b \\
\hline
e & e & a & b \\
a & a &   & e \\
b & b & e &
\end{array}
\tag{2.24}
$$

Now we know, by Problem 2.16, that each row and column of the array

23

must, at least, contain different elements of G, thus we obtain finally

$$
\begin{array}{c|ccc}
 & e & a & b \\
\hline
e & e & a & b \\
a & a & b & e \\
b & b & e & a
\end{array}
\tag{2.25}
$$

(iv) Let $G = \{e, a, b, c\}$. We distinguish two cases.

Case 1. Suppose there exists an element whose inverse is not equal to itself. For the sake of argument suppose $ab = e$. Then $ba = e$ and hence $ac \neq e, a$ or c. Thus $ac = b$ and similarly $ca = b$. Hence

$$b^2 = bb = b(ac) = (ba)c = ec = c.$$

We obtain the array

$$
\begin{array}{c|cccc}
 & e & a & b & c \\
\hline
e & e & a & b & c \\
a & a & c & e & b \\
b & b & e & & c \\
c & c & b & &
\end{array}
\tag{2.26}
$$

which we complete to yield

$$
\begin{array}{c|cccc}
 & e & a & b & c \\
\hline
e & e & a & b & c \\
a & a & c & e & b \\
b & b & e & c & a \\
c & c & b & a & e
\end{array}
\tag{2.27}
$$

Case 2. Suppose now every element has inverse equal to itself. Thus $a^2 = b^2 = c^2 = e$. Then $ab \neq e, a$ or b and so $ab = c$. Similarly $ba = c$. by similar argument we have $ac = ca = b$, $ac = cb = a$, yielding

$$
\begin{array}{c|cccc}
 & e & a & b & c \\
\hline
e & e & a & b & c \\
a & b & e & c & d \\
b & b & c & e & a \\
c & c & b & a & e
\end{array}
\tag{2.28}
$$

This last group is called the Klein four-group (after F. Klein, German, 1849–1925). $\qquad\square$

24

Although we have obtained the above tables in a formal manner (not strictly guaranteeing the existence of the groups) we find that we have concrete realisations of the above, namely

(i) $G = \{1\}$

(ii) $G = \{1, -1\}$

(iii) $G = \left\{1, \dfrac{-1+i\sqrt{3}}{2}, \dfrac{-1-i\sqrt{3}}{2}\right\}$

(iv) *Case* 1: $G = \{1, i, -i, -1\}$

Case 2:
$$G = \left\{\begin{pmatrix} 1 & 0 \\ 0 & 1 \end{pmatrix}, \begin{pmatrix} -1 & 0 \\ 0 & -1 \end{pmatrix}, \begin{pmatrix} 0 & 1 \\ 1 & 0 \end{pmatrix}, \begin{pmatrix} 0 & -1 \\ -1 & 0 \end{pmatrix}\right\}$$

the obvious laws of multiplication being employed. □

Problem 2.20 Prove that the set G of nonsingular 2×2 matrices over the rational numbers $\mathbb{Q}$ forms a group which is not commutative. Find $\mathbf{a}, \mathbf{b} \in G$ such that $(\mathbf{ab})^{-1} \neq \mathbf{a}^{-1}\mathbf{b}^{-1}$.

Solution. It is not stated how the matrices are to be multiplied but it is customary to assume that ordinary matrix multiplication is intended, this multiplication being associative and the identity 2×2 matrix being the identity for G.

Let $\mathbf{x}, \mathbf{y}$ be two nonsingular 2×2 matrices over $\mathbb{Q}$, say

$$\mathbf{x} = \begin{pmatrix} x_{11} & x_{12} \\ x_{21} & x_{22} \end{pmatrix}, \qquad \mathbf{y} = \begin{pmatrix} y_{11} & y_{12} \\ y_{21} & y_{22} \end{pmatrix}$$

where the matrix entries are rational numbers and where

$$x_{11}x_{22} - x_{12}x_{21} \neq 0, \qquad y_{11}y_{22} - y_{21}y_{12} \neq 0.$$

Then $\mathbf{xy}$ is also nonsingular for

$$\mathbf{xy} = \begin{pmatrix} x_{11}y_{11} + x_{12}y_{21} & x_{11}y_{12} + x_{12}y_{22} \\ x_{21}y_{11} + x_{22}y_{21} & x_{21}y_{12} + x_{22}y_{22} \end{pmatrix}$$

and
$$(x_{11}y_{11} + x_{12}y_{21})(x_{21}y_{12} + x_{22}y_{22}) - (x_{21}y_{11} + x_{22}y_{21})(x_{11}y_{12} + x_{12}y_{22})$$
$$= (x_{11}x_{22} - x_{21}x_{12})(y_{11}y_{22} - y_{21}y_{12}) \neq 0.$$

Also, if $\mathbf{z} \in G$, say

$$\mathbf{z} = \begin{pmatrix} z_{11} & z_{12} \\ z_{21} & z_{22} \end{pmatrix}$$

then we deduce that $z^{-1} \in G$ is

$$z^{-1} = \frac{1}{z_{11}z_{22} - z_{12}z_{21}} \begin{pmatrix} z_{22} & -z_{12} \\ -z_{21} & z_{11} \end{pmatrix}.$$

Thus we have established that G is a group. To show that G is not commutative it is sufficient to observe that if we let

$$\mathbf{a} = \begin{pmatrix} 1 & 1 \\ 0 & 1 \end{pmatrix}, \qquad \mathbf{b} = \begin{pmatrix} 8 & 5 \\ 3 & 2 \end{pmatrix}$$

then

$$\mathbf{ab} = \begin{pmatrix} 11 & 7 \\ 3 & 2 \end{pmatrix} \neq \begin{pmatrix} 8 & 13 \\ 3 & 5 \end{pmatrix} = \mathbf{ba}.$$

We also note that

$$\mathbf{a}^{-1}\mathbf{b}^{-1} = \begin{pmatrix} 1 & -1 \\ 0 & 1 \end{pmatrix}\begin{pmatrix} 2 & -5 \\ -3 & 8 \end{pmatrix} = \begin{pmatrix} 5 & -13 \\ -3 & 8 \end{pmatrix}$$

$$\neq \begin{pmatrix} 2 & -7 \\ -3 & 11 \end{pmatrix} = (\mathbf{ab})^{-1}. \qquad \square$$

We may generalise the above problem to the case of nonsingular $n \times n$ matrices. We merely require to know the product rule for determinants, namely $\det(\mathbf{x}) \det(\mathbf{y}) = \det(\mathbf{xy})$.

Problem 2.21 Let $x, y \in G$, where G is a group. Prove that

$$(xy)^{-1} = y^{-1}x^{-1}. \tag{2.29}$$

Solution.

$$(xy)(y^{-1}x^{-1}) = x(yy^{-1})x^{-1} = xex^{-1} = xx^{-1} = e.$$

Hence

$$(xy)^{-1} = y^{-1}x^{-1}. \qquad \square$$

This reversal rule for the inverse of a product of two elements extends to the inverse of the product of three elements, thus if $z \in G$ then

$$(xyz)^{-1} = z^{-1}y^{-1}x^{-1}. \tag{2.30}$$

There is an obvious extension to the product of n elements

$$(x_1 x_2 \ldots x_n)^{-1} = x_n^{-1} x_{n-1}^{-1} \ldots x_1^{-1} \quad (x_i \in G, \quad i = 1, 2, \ldots, n). \tag{2.31}$$

On letting $x_1 = x_2 = \ldots = x_n = x$ in equation 2.31 we have

$$(x^n)^{-1} = (x^{-1})^n \quad (n = 1, 2, \ldots). \tag{2.32}$$

Conventionally we let $x^0 = e$ and then in G the general index law (cf. equation 2.13) holds, namely

$$x^m x^n = x^{m+n} \quad (m, n \in \mathbb{Z}). \tag{2.33}$$

A commutative group is often called an *Abelian* group (after N. H. Abel, Norwegian, 1802–1829).

Problem 2.22   Let G be a group in which $x^2 = e$ for all $x \in G$. Prove that G is Abelian.

Solution. For all $x \in G$, since $x^2 = e$ we have $x^{-1} = x$. Hence for all $a, b \in G$ we have
$$ab = (ab)^{-1} = b^{-1}a^{-1} = ba. \qquad \square$$

A group is said to be *cyclic* if every element is a power of some given element which is said to *generate*, or to be a *generator* of, the group. A cyclic group is necessarily Abelian.

Problem 2.23   Which of the five groups in Problem 2.19 are cyclic?

Solution.

(i) G is cyclic, a generator is e.
(ii) G is cyclic, a generator is a but not e.
(iii) G is cyclic, a generator is a or b but not e.
(iv) G of *Case* 1 is cyclic, a generator is a or b but not e or c.
      G of *Case* 2 is not cyclic. $\qquad \square$

If G has a finite number of elements then G is said to have *finite order*, the precise number of elements being called the *order* of G written $|G|$; otherwise we say G has *infinite order*. If $x \in G$ and there exists $n \geqslant 1$ such that $x^n = e$ then x is said to have *finite order*, otherwise x has *infinite order*; if $n = 1$ then $x = e$ and x is said to have order 1, if $n > 1$ and if $x^m \neq e\,(1 < m < n)$ then x is said to have order n. It is true, but not entirely obvious, that x has finite order if and only if the (sub-)group generated by x has finite order. In a finite group every element has finite order but the converse is false.

Problem 2.24   Give an example of a group in which every nonidentity element has infinite order.

Solution. Let $G = \{2^r : r = 0, \pm 1, \ldots\}$ where the usual multiplication in $\mathbb{Q}$ is intended. It is clear that G is a group and that no nonidentity element has finite order since $(2^r)^n = 1$ if and only if $2^{rn} = 1$ if and only if $r = 0$ or $n = 0$.

G is an example of an infinite cyclic group. $\qquad \square$

Problem 2.25   Give an example of an infinite group in which every element has finite order.

c

Solution. Let

$$G = \left\{ \cos\frac{2k\pi}{3^n} + i\sin\frac{2k\pi}{3^n} : n = 0, 1, 2, \ldots; \ k = 0, \pm 1, \pm 2 \ldots \right\}$$

where the usual multiplication of the complex numbers $\mathbb{C}$ is intended. To verify that G is a group is easy, the only nontrivial part of the verification being to show that multiplication in G is closed and this follows since

$$\left(\cos\frac{2k_1\pi}{3^{n_1}} + i\sin\frac{2k_1\pi}{3^{n_1}} \right)\left(\cos\frac{2k_2\pi}{3^{n_2}} + i\sin\frac{2k_2\pi}{3^{n_2}} \right)$$

$$= \cos\left(\frac{2k_1\pi}{3^{n_1}} + \frac{2k_2\pi}{3^{n_2}}\right) + i\sin\left(\frac{2k_1\pi}{3^{n_1}} + \frac{2k_2\pi}{3^{n_2}}\right)$$

$$= \cos\left\{\frac{2(3^{n_2}k_1 + 3^{n_1}k_2)\pi}{3^{n_1+n_2}}\right\} + i\sin\left\{\frac{2(3^{n_2}k_1 + 3^{n_1}k_2)\pi}{3^{n_1+n_2}}\right\}.$$

Clearly G is an infinite group but every element of G has finite order for, by Demoivre's theorem,

$$\left(\cos\frac{2k\pi}{3^n} + i\sin\frac{2k\pi}{3^n} \right)^{3^n} = \cos 2k\pi + i\sin 2k\pi$$

$$= 1.$$

In fact G is the group of all complex 3^n-roots of 1 ($n = 0, 1, 2, \ldots$). ☐

Problem 2.26 Prove that an element x of a group G has order n if and only if $\{x^s : s = 1, 2, \ldots\}$ consists of precisely n elements.

Solution. If $n = 1$ there is nothing to prove. Suppose that x has order $n > 1$. Then the n elements $x, x^2, \ldots, x^{n-1}, x^n (= e)$ are distinct for if $x^u = x^v$ where $1 \leqslant u < v \leqslant n$ then

$$x^{v-u} = x^v x^{-u} = x^v(x^u)^{-1}$$

$$= x^v(x^v)^{-1} = x^{v-v} = x^0 = e$$

and, since $1 \leqslant u-v < n$, we obtain a contradiction. It will follow that $\{x^s : s = 1, 2, \ldots\}$ consists of precisely n elements if we can show that for any power x^a of x we have $x^a \in \{x, x^2, \ldots, x^n\}$. By the division algorithm for the rational integers, there exist $q, r \in \mathbb{Z}$ such that

$$a = nq + r \quad (0 \leqslant r < n)$$

and hence

$$x^a = x^{nq+r} = (x^n)^q x^r = ex^r = x^r \text{ from which } x^a \in \{x, x^2, \ldots, x^n\}.$$

Conversely suppose $\{x^s : s = 1, 2, \ldots\}$ consists of precisely n elements and that x has order m. Then, by the first part of the solution $\{x^s : s = 1, 2, \ldots\}$ consists of precisely m elements and so $m = n$. ☐

It follows from Problem 2.26 that if G is a cyclic group generated by x then $|G|$ is equal to the order of x.

Problem 2.27   Let G be the six-element group $\{e, a, b, c, d, f\}$ with multiplication:

	e	a	b	c	d	f
e	e	a	b	c	d	f
a	a	b	e	d	f	c
b	b	e	a	f	c	d
c	c	f	d	e	b	a
d	d	c	f	a	e	b
f	f	d	c	b	a	e

$$(2.34)$$

Find the orders of the elements of G.

Solution. (Notice that we are *assuming* that the above array does represent a group.) e has order 1.

$a^2 = b$, $a^3 = ab = e$ implies a has order 3.
$b^2 = a$, $b^3 = ab = e$ implies b has order 3.
$c^2 = e$, $d^2 = e$, $f^2 = e$ implies c, d, f have order 2. $\square$

Problem 2.28   Let a, x belong to the group G and let $b = x^{-1}ax$. Prove that a and b have the same order.

Solution. We observe first that if $n \in \mathbb{Z}$ then
$$b^n = x^{-1}a^n x \qquad (2.35)$$
for

$$
\begin{aligned}
b^n &= (x^{-1}ax)^n \\
&= (x^{-1}ax)(x^{-1}ax)\dots(x^{-1}ax) \\
&= x^{-1}a(xx^{-1})a(xx^{-1})\dots(xx^{-1})ax \\
&= x^{-1}aeae\dots eax \\
&= x^{-1}aa\dots ax \\
&= x^{-1}a^n x.
\end{aligned}
$$

Thus if $a^n = e$ then $b^n = e$ and conversely.

We say that $b \in G$ is *conjugate* to $a \in G$ whenever there exists $x \in G$ such that $b = x^{-1}ax$. $\square$

Problem 2.29   Prove that the relation 'is conjugate to' is an equivalence relation on the group G.

Solution. If $a \in G$ then a is conjugate to a for $a = e^{-1}ae$. If b is conjugate to a then $b = x^{-1}ax(x \in G)$ and

$$a = (xx^{-1})a(xx^{-1})$$
$$= x(x^{-1}ax)x^{-1}$$
$$= xbx^{-1}$$
$$= (x^{-1})^{-1}b(x^{-1})$$

and so a is conjugate to b. If b is conjugate to a and c is conjugate to b then there exist $x, y \in G$ such that $b = x^{-1}ax$, $c = y^{-1}by$ and so

$$c = y^{-1}by = y^{-1}(x^{-1}ax)y$$
$$= y^{-1}x^{-1}axy$$
$$= (xy)^{-1}a(xy).$$

Thus c is conjugate to a. We have now established that the relation is reflexive, symmetric and transitive giving the result. $\qquad \square$

We speak now of elements being *conjugate* and the equivalence classes of being *conjugacy classes*. Thus the conjugacy class containing a is $\{t^{-1}at : t \in G\}$. An element is called *self-conjugate* or *central* if its conjugacy class consists of itself alone, i.e. a is central if $t^{-1}at = a$ (for all $t \in G$). The set of self-conjugate elements of G is called the *centre* $Z(G)$ of G.

Problem 2.30 Find the conjugacy classes in the group G of Problem 2.27.

Solution. $\{e\}$ is a class of one element.

$$\{x^{-1}ax : x \in G\} = \{e^{-1}ae, a^{-1}aa, b^{-1}ab, c^{-1}ac, d^{-1}ad, f^{-1}af\}$$
$$= \{a, a, a, b, b, b\}$$
$$= \{a, b\}$$
$$\{x^{-1}cx : x \in G\} = \{e^{-1}ce, a^{-1}ca, b^{-1}cb, c^{-1}cc, d^{-1}cd, f^{-1}cf\}$$
$$= \{c, d, f, c, f, d\} = \{c, d, f\}.$$

Thus G has the three conjugacy classes $\{e\}$, $\{a, b\}$, $\{c, d, f\}$. $\qquad \square$

2.3 Subgroups

Problem 2.31 Let H be a nonempty subset of the group G such that for all $x, y \in H$, we have $xy \in H$ and $x^{-1} \in H$. Prove that H is a group having the same law of composition as G.

Solution. By assumption H is closed under multiplication. This multiplication, being associative in G, is certainly associative in H. Let $a \in H$, then $a^{-1} \in H$ and so $e = aa^{-1} \in$ H. Thus H is a monoid in which every element has an inverse, i.e. H is a group.

H is called a *subgroup* of G; we note that e is also the identity of H. $\qquad \square$

The group G is called *nontrivial* if $G \neq \{e\}$. A nontrivial group has at least two subgroups namely G and $\{e\}$, any other subgroup is called a *proper* subgroup.

Problem 2.32 Let a be an element of the group G and let $C_G(a) = \{x \in G : x^{-1}ax = a\}$. Prove that $C_G(a)$ is a subgroup of G.

Solution. We verify first that $C_G(a) \neq \varnothing$. This is clear since $e \in C_G(a)$. Let $x, y \in C_G(a)$. Then

$$
\begin{aligned}
(xy)^{-1}a(xy) &= (y^{-1}x^{-1})a(xy) \\
&= y^{-1}(x^{-1}ax)y \\
&= y^{-1}ay \\
&= a
\end{aligned}
$$

$$
\begin{aligned}
\text{and } (x^{-1})^{-1}a(x^{-1}) &= xax^{-1} \\
&= x(x^{-1}ax)x^{-1} \\
&= (xx^{-1})a(xx^{-1}) \\
&= eae \\
&= a.
\end{aligned}
$$

Thus $xy, x^{-1} \in C_G(a)$ and so we conclude that $C_G(a)$ is a subgroup. $\square$

We observe that $x^{-1}ax = a$ if and only if $ax = xa$ and so $C_G(a)$ is the set of all elements of G *commuting* with a. We call $C_G(a)$ the *centraliser* of a.

Problem 2.33 Prove that the centre $Z(G)$ is a subgroup of the group G.

Solution. By definition

$$Z(G) = \{z \in G : z^{-1}az = a \text{ for all } a \in G\}.$$

Thus letting $x, y \in Z(G)$ we see that $x, y \in C_G(a)$ for all $a \in G$. Hence $xy, x^{-1} \in C_G(a)$ for all $a \in G$ and so $xy, x^{-1} \in Z(G)$. Thus $Z(G)$ is a subgroup. $\square$

Implicit in the above solution is the relation

$$Z(G) = \bigcap_{a \in G} C_G(a). \tag{2.36}$$

Problem 2.34 Find the centre of the group $GL(2, \mathbb{R})$ of nonsingular 2×2 matrices over the real numbers.

Solution.

$$\text{Let } \mathbf{z} = \begin{pmatrix} \alpha & \beta \\ \gamma & \delta \end{pmatrix}$$

31

belong to the centre of $GL(2, \mathbb{R})$. Then $\mathbf{z}$ commutes with all nonsingular 2×2 matrices and so, in particular, $\mathbf{z}$ commutes with

$$\begin{pmatrix} 1 & 1 \\ 0 & 1 \end{pmatrix}, \qquad \begin{pmatrix} 1 & 0 \\ 1 & 1 \end{pmatrix}.$$

But

$$\begin{aligned} \begin{pmatrix} \alpha & \alpha+\beta \\ \gamma & \gamma+\delta \end{pmatrix} &= \begin{pmatrix} \alpha & \beta \\ \gamma & \delta \end{pmatrix}\begin{pmatrix} 1 & 1 \\ 0 & 1 \end{pmatrix} \\ &= \begin{pmatrix} 1 & 1 \\ 0 & 1 \end{pmatrix}\begin{pmatrix} \alpha & \beta \\ \gamma & \delta \end{pmatrix} \\ &= \begin{pmatrix} \alpha+\gamma & \beta+\delta \\ \gamma & \delta \end{pmatrix} \end{aligned} \qquad (2.37)$$

and so we must have $\gamma = 0$ and $\alpha = \delta$. Repeating the argument with $\begin{pmatrix} 1 & 0 \\ 1 & 1 \end{pmatrix}$ we have $\beta = 0$. Thus

$$\mathbf{z} = \begin{pmatrix} \alpha & 0 \\ 0 & \alpha \end{pmatrix}$$

where, necessarily, $\alpha \neq 0$. But $\mathbf{z}$ is now a scalar matrix and so commutes with all 2×2 matrices (nonsingular or not). Hence the centre of $GL(2, \mathbb{R})$ consists of all nonzero scalar matrices.   □

$GL(2, \mathbb{R})$ is called the *general linear group* of 2×2 matrices. The non-singular $n \times n$ matrices over $\mathbb{R}$ form the *general linear group* $GL(n, \mathbb{R})$ the centre of which again consists of the nonzero scalar matrices.

Problem 2.35 Let A and B be subgroups of the group G. Prove that $A \cap B$ is a subgroup of G.

Solution. First we observe that $A \cap B \neq \varnothing$ for $e \in A$ and $e \in B$. Let $x, y \in A \cap B$. Then $x, y \in A$ and so $xy, x^{-1} \in A$, similarly $xy, x^{-1} \in B$. Thus $xy, x^{-1} \in A \cap B$ giving the result (Problem 2.31).   □

If $A_1, A_2, \ldots, A_n$ are n subgroups of G then $A_1 \cap A_2 \cap \ldots \cap A_n$ is a subgroup of G. Indeed if $H_\lambda(\lambda \in \Lambda)$ are subgroups of G, Λ being some indexing set, then $\bigcap_{\lambda \in \Lambda} H_\lambda$ is a subgroup of G. We have already characterised the centre $Z(G)$ of G as

$$Z(G) = \bigcap_{a \in G} C_G(a).$$

Problem 2.36 Let H be a subgroup of the group G and let $a \in G$. Prove that $a^{-1}Ha = \{a^{-1}ha : h \in H\}$ is a subgroup of G.

Solution. We see that $a^{-1}Ha \neq \varnothing$ for $e = a^{-1}ea \in a^{-1}Ha$. Let

$x, y \in a^{-1}Ha$. Then there exist $r, s \in H$ such that $x = a^{-1}ra$, $y = a^{-1}sa$. Then

$$\begin{aligned}
xy &= (a^{-1}ra)(a^{-1}sa) \\
&= a^{-1}r(aa^{-1})sa \\
&= a^{-1}resa \\
&= a^{-1}(rs)a \in a^{-1}Ha \\
x^{-1} &= a^{-1}r^{-1}a \in a^{-1}Ha,
\end{aligned}$$

$a^{-1}Ha$ is called a *conjugate* of H in G. $\qquad\square$

A subgroup that coincides with all its conjugates in G is said to be *self-conjugate* or *normal* in G. In an Abelian group every subgroup is normal.

Problem 2.37 Prove that $Z(G)$ is normal in G.

Solution. Let $x \in G$. Then for all $z \in Z(G)$, $x^{-1}zx = z$ and so

$$\begin{aligned}
x^{-1}Z(G)x &= \{x^{-1}zx : z \in Z(G)\} \\
&= \{z : z \in Z(G)\} \\
&= Z(G). \qquad\square
\end{aligned}$$

Problem 2.38 Find six subgroups in the group G given in Problem 2.27. Which are normal and what is the centre?

Solution. Obviously $\{e\}$ and G are subgroups but unfortunately no very systematic method exists for finding subgroups of a group. By inspection of (2.34) we determine four subgroups $\{e, a, b\}$, $\{e, c\}$, $\{e, d\}$, $\{e, f\}$. $\{e, a, b\}$ is normal for

$$\begin{aligned}
e^{-1}\{e, a, b\}e &= \{e, a, b\} \\
a^{-1}\{e, a, b\}a &= \{a^{-1}ea, a^{-1}aa, a^{-1}ba\} \\
&= \{e, a, b\} \\
b^{-1}\{e, a, b\}b &= \{b^{-1}eb, b^{-1}ab, b^{-1}bb\} \\
&= \{e, a, b\} \\
c^{-1}\{e, a, b\}c &= \{c^{-1}ec, c^{-1}ac, c^{-1}bc\} \\
&= \{e, b, a\} \\
d^{-1}\{e, a, b\}d &= \{d^{-1}ed, d^{-1}ad, d^{-1}bd\} \\
&= \{e, b, a\} \\
f^{-1}\{e, a, b\}f &= \{f^{-1}ef, f^{-1}af, f^{-1}bf\} \\
&= \{e, b, a\}.
\end{aligned}$$

The three subgroups of order 2 are conjugate to one another, for example

$$a^{-1}\{e,c\}a = \{e,a^{-1}ca\} = \{e,d\}$$
$$d^{-1}\{e,c\}d = \{e,d^{-1}cd\} = \{e,f\}$$
$$a^{-1}\{e,d\}a = \{e,a^{-1}da\} = \{e,f\}.$$

In this instance $Z(G) = \{e\}$. ☐

Problem 2.39 Let H be a subgroup of G such that $x^{-1}Hx \subseteq H$ for all $x \in G$. Prove that H is normal in G.

Solution. Let $y \in G$, we must show that $y^{-1}Hy = H$. Let $z = y^{-1}$ then certainly $z^{-1}Hz \subseteq H$ and hence

$$H = z(z^{-1}Hz)z^{-1} \subseteq zHz^{-1} = y^{-1}Hy \subseteq H. \qquad ☐$$

Problem 2.40 Let A and B be normal subgroups of the group G. Prove that $A \cap B$ is a normal subgroup of G.

Solution. We already know that $A \cap B$ is a subgroup of G. Let $x \in G$, then, since A is normal,

$$x^{-1}(A \cap B)x \subseteq x^{-1}Ax \subseteq A.$$

Similarly

$$x^{-1}(A \cap B)x \subseteq B.$$

Hence

$$x^{-1}(A \cap B)x \subseteq A \cap B. \qquad ☐$$

Problem 2.41 Let H be a subgroup of the group G and let $N_G(H) = \{n \in G : n^{-1}Hn = H\}$. Prove that $N_G(H)$ is a subgroup of G containing H.

Solution. Since $h^{-1}Hh = H$ for all $h \in H$, $H \subseteq N_G(H)$. Let $x, y \in N_G(H)$, then

$$(xy)^{-1}H(xy) = (y^{-1}x^{-1})H(xy)$$
$$= y^{-1}(x^{-1}Hx)y$$
$$= y^{-1}Hy$$
$$= H$$
$$(x^{-1})^{-1}H(x^{-1}) = xHx^{-1}$$
$$= x(x^{-1}Hx)x^{-1}$$
$$= (xx^{-1})H(xx^{-1}).$$

$N_G(H)$ is called the *normaliser* of H in G. $N_G(H) = G$ if and only if H is normal in G. ☐

Problem 2.42 Let G be a finite cyclic group of order n. Let d be a positive divisor of n. Prove that G has a subgroup of order d.

Solution. If $d = 1$ or n there is nothing to prove. Thus we suppose $1 < d < n$. Since G is cyclic of order n there exists $x \in G$ such that the n elements of G are precisely the elements $e, x, x^2, \ldots, x^{n-1}$. Let $n = dm$ ($m \in \mathbb{Z}$) and let $H = \{x^{sm} : s = 0, \pm 1, \ldots\}$. Then as

$$x^{sm} x^{tm} = x^{(s+t)m}$$
$$(x^{sm})^{-1} = x^{(-s)m}$$

it follows that H is a subgroup. To determine $|H|$ we observe that, by the division algorithm, there exist $q, r \in \mathbb{Z}$ such that

$$s = dq + r \qquad (0 \leqslant r < d)$$

and so

$$x^{sm} = x^{(dq+r)m}$$
$$= (x^{dm})^q x^{rm}$$
$$= (x^n)^q x^{rm}$$
$$= x^{rm}$$

from which we deduce that

$$H = \{x^{rm} : r = 0, 1, \ldots, d-1\}.$$

Furthermore the elements $e, x^m, x^{2m}, \ldots, x^{(d-1)m}$ are distinct. Thus $|H| = d$.

It may be shown that if K is another subgroup such that $|K| = d$ then $K = H$. As we shall see, the order of any subgroup must divide the order of the group and so we can conclude that every subgroup of a cyclic group is cyclic and that there exists a unique subgroup for every divisor of the order of the group. $\qquad\square$

Problem 2.43 Let G be a group having no proper subgroups. Prove that $|G| = 1$ or p where p is a prime.

Solution. If $|G| = 1$ there is nothing to prove. Let $|G| > 1$ and let $x \in G$, $x \neq e$. Then $H = \{x^t : t = 0, \pm 1, \ldots\}$ is easily shown to be a subgroup of G and so $G = H$. Now $K = \{x^{2t} : t = 0, \pm 1, \ldots\}$ is also a subgroup. If $x^2 = e$ then $K = \{e\}$ and $G = \{e, x\}$ has order 2. If $x^2 \neq e$ then $K \neq \{e\}$ and so $K = G$. Thus $x \in K$ and so there exists t_0 such that $x = x^{2t_0}$ from which we conclude that $x^{2t_0-1} = e$, so that x and G have finite order. Thus G is a finite cyclic group. Let p be a prime divisor of n. Then G has a subgroup of order p and so $|G| = n = p$. $\qquad\square$

If X, Y are nonempty subsets of G then XY denotes the subset

$$XY = \{xy : x \in X, y \in Y\}. \tag{2.38}$$

If Z is a third nonempty subset of G then the associativity of the law of composition of elements in G implies that

$$(XY)Z = X(YZ). \tag{2.39}$$

Further if we have the inclusion relation $X \subseteq Y$ then we have $XZ \subseteq XY$ and $ZX \subseteq ZY$. If, in particular, X has only one element x_0 it is usual to omit brackets and to write

$$XY = \{x_0\}Y = x_0Y = \{x_0y : y \in Y\}. \tag{2.40}$$

Similarly if $Y = \{y_0\}$ we have

$$XY = X\{y_0\} = Xy_0 = \{xy_0 : x \in X\}. \tag{2.41}$$

These definitions extend to the product of any finite number of nonempty subsets of G and are consistent with the earlier definition that $a^{-1}Ha = \{a^{-1}ha : h \in H\}$. $\qquad\square$

Problem 2.44 Let H and N be subgroups of the group G. Prove that HN is a subgroup if and only if $HN = NH$.

Solution. Suppose HN is a subgroup. Then H and N are subsets of HN and so by the closure of the multiplication in HN, $NH \subseteq HN$. We now wish to show $HN \subseteq NH$. Let $x \in HN$, then $x^{-1} \in HN$ and so $x^{-1} = hn$ for some $n \in N$, $h \in H$. Thus $x = (hn)^{-1} = n^{-1}h^{-1} \in NH$. Hence $HN = NH$.

Conversely if $HN = NH$ let $h_1 n_1$ and $h_2 n_2$ be typical elements of HN ($h_i \in H$, $n_i \in N$; $i = 1, 2$). Then $n_1 h_2 \in NH = HN$ and so there exist $h_3 \in H$, $n_3 \in N$ such that $n_1 h_2 = h_3 n_3$. Hence we have

$$
\begin{aligned}
(h_1 n_1)(h_2 n_2) &= h_1(n_1 h_2)n_2 \\
&= h_1(h_3 n_3)n_2 \\
&= (h_1 h_3)(n_3 n_2) \in HN
\end{aligned}
$$
$$(h_1 n_1)^{-1} = n_1^{-1} h_1^{-1} \in NH = HN.$$

Thus HN is a subgroup. $\qquad\square$

Problem 2.45 Let H and N be subgroups of the group G and let N be normal in G. Prove that HN is a subgroup of G.

Solution. Since N is normal in G we have

$$Nx = xx^{-1}Nx = xN \quad \text{(for all } x \in G).$$

Hence

$$HN = \bigcup_{h \in H} hN = \bigcup_{h \in H} Nh = NH. \qquad\square$$

If H is a subgroup of the group G and if $g \in G$ then $g \in gH$ where

$$gH = \{ gh : h \in H \} \tag{2.42}$$

and we call gH the *left coset* of H in G determined by g. Similarly $g \in Hg$ where

$$Hg = \{hg : h \in H\} \tag{2.43}$$

which is called the *right* coset of H in G determined by g. We remark that gH (or Hg) equals H if and only if $g \in H$.

Problem 2.46 Let x, y be elements of the group G and let H be a subgroup of G. Prove that the following statements are equivalent:
(i) $Hx = Hy$.
(ii) There exists $h \in H$ such that $y = hx$.
(iii) $xy^{-1} \in H$.

Solution. (i) implies (ii). We have $y \in Hy = Hx$ and so there exists $h \in H$ such that $y = hx$. (ii) implies (iii). If $y = hx\,(h \in H)$ then
$$xy^{-1} = x(hx)^{-1} = x(x^{-1}h^{-1})$$
$$= (xx^{-1})h^{-1} = eh^{-1} \in H.$$
(iii) implies (i). If $xy^{-1} \in H$ then $H(xy^{-1}) = H$ and hence
$$Hy = [H(xy^{-1})]y = H[(xy^{-1})y]$$
$$= H[x(y^{-1}y)] = Hx. \qquad \square$$

Problem 2.47 Prove that the relation defined on the group G by 'x determines the same right coset as y' $(x, y \in G)$ is an equivalence relation on G.

Solution. Clearly $Hx = Hx$ and $Hy = Hz$ implies $Hz = Hy$, thus the relation is reflexive and symmetric. Transitivity is also immediate for $Ha = Hb$ and $Hb = Hc$ implies $Ha = Hc$ $(a, b, c \in G)$. $\qquad \square$

Since the above relation is an equivalence relation, G is the disjoint union of the equivalence classes which are the right cosets. G is also the disjoint union of its left cosets.

Problem 2.48 Let G be the group in Problem 2.27 and let $H = \{e, c\}$. Write G as a disjoint union of right cosets of H in G.

Solution. $H = He$ is one such coset. We choose an element a (say), not in H, then
$$Ha = \{ea, ca\} = \{a, f\}.$$
We now choose an element which is not in $H \cup Ha$, b say,
$$Hb = \{eb, cb\} = \{b, d\}.$$
Thus
$$G = H \cup Ha \cup Hb$$
is a representation of G as a disjoint union of right cosets.

A representation of a group as a disjoint union of right (or left) cosets is called a *coset decomposition*. $\qquad \square$

Problem 2.49 Let G be a finite group and let H be a subgroup of G. Prove that $|H|$ divides $|G|$ and that $|G|/|H|$ equals the number of distinct right (or left) cosets of H in G.

Solution. Since $h_1 g = h_2 g$ if and only if $h_1 = h_2$ ($g \in G$, $h_1 h_2 \in H$) it follows that the number of elements in any right coset is exactly equal to $|H|$. Suppose we have n distinct right cosets. Then these right cosets are mutually disjoint and so each element of G appears once and only once in their union. Thus $n|H| = |G|$. Hence also the numbers of left and right cosets are equal to $|G|/|H|$.

This very important discovery is due to J. L. Lagrange (French, 1736–1813). $\square$

If G is finite then we write $|G : H|$ for the number of left (or right) cosets of H in G and we have then

$$|G| = |G : H||H|. \tag{2.44}$$

We call $|G : H|$ the *index* of H in G. If G is infinite then the numbers of distinct left and right cosets of a subgroup H of G are either both finite or both infinite. We again define the *index* $|G : H|$ as the number (possibly infinite) of distinct right cosets.

Problem 2.50 Let H be a subgroup of the group G. Prove that $Hg = gH$ for all $g \in G$ if and only if H is a normal subgroup of G.

Solution. Suppose H is a normal subgroup of G. Then for all $g \in G$, $Hg = (gg^{-1})(Hg) = g(g^{-1}Hg) = gH$. Conversely if $Hg = gH$ for all $g \in G$ we have $g^{-1}Hg = g^{-1}(gH) = (g^{-1}g)H = H$ and so H is normal in G. $\square$

Problem 2.51 Let H be a subgroup of the group G of index 2. Prove that H is normal in G.

Solution. We are given that H has precisely two distinct cosets in G. Now if $x \in G$, $x \notin H$ then the cosets H, xH are distinct and so $G = H \cup xH$ where $H \cap xH = \varnothing$. Similarly we have $G = H \cup Hx$ where $H \cap Hx = \varnothing$. Thus, by equation 1.8, $xH = xH \cap G = xH \cap (H \cup Hx) = (xH \cap H) \cup (xH \cap Hx) = xH \cap Hx$ and so $Hx = xH \cap Hx = xH$. Thus, for all $x \in G$, $x \notin H$, $xH = Hx$. Since, trivially, $yH = H = Hy$ for all $y \in H$ we have $zH = Hz$ for all $z \in G$. Thus H is normal in G. $\square$

Problem 2.52 Prove that if the group G has finite order then the order of every element of G divides $|G|$.

Solution. Let $x \in G$ and suppose x has order d. Then x generates a subgroup of order d. Hence d divides $|G|$ (Problem 2.49). $\square$

Problem 2.53 Let a be an element of the group G such that the centraliser $C_G(a)$ has finite index n in G. Prove that a has precisely n conjugates in G.

Solution. Let $x_1, x_2, \ldots, x_n \in G$ be such that
$$G = C_G(a)x_1 \cup C_G(a)x_2 \cup \ldots \cup C_G(a)x_n$$
is a coset decomposition of $C_G(a)$ in G.

We establish the result by showing that the conjugates of a, $x_1^{-1}ax_1$, $x_2^{-1}ax_2, \ldots, x_n^{-1}ax_n$, are distinct and that a has no other conjugate. We observe first that $x_i^{-1}ax_i = x_j^{-1}ax_j$ implies $x_j x_i^{-1}ax_i x_j^{-1} = a$ and thus $(x_i x_j^{-1})^{-1}a(x_i x_j^{-1}) = a$. From this we deduce that $x_i x_j^{-1} \in C_G(a)$ and so $C_G(a)x_i = C_G(a)x_j$. This implies $i = j$, and so the n listed conjugates of a are distinct. Second, if b is a conjugate of a then $b = x^{-1}ax$ $(x \in G)$. Now x must belong to some coset of $C_G(a)$ in G, suppose $x \in C_G(a)x_k$. Then $x = cx_k$ $(c \in C_G(a))$ and hence
$$\begin{aligned} b = x^{-1}ax &= (cx_k)^{-1}a(cx_k) \\ &= x_k^{-1}(c^{-1}ac)x_k \\ &= x_k^{-1}ax_k. \end{aligned} \qquad \square$$

In a finite group the number of conjugates of an element, being the index of the centraliser of the element, divides the order of the group.

Problem 2.54 Let a, x be elements of the group G and let $b = x^{-1}ax$. Prove that $C_G(b) = x^{-1}C_G(a)x$.

Solution. Let $y \in C_G(a)$. Then
$$\begin{aligned} (x^{-1}yx)^{-1}b(x^{-1}yx) &= (x^{-1}y^{-1}x)(x^{-1}ax)(x^{-1}yx) \\ &= x^{-1}y^{-1}(xx^{-1})a(xx^{-1})yx \\ &= x^{-1}(y^{-1}ay)x \\ &= x^{-1}ax \\ &= b. \end{aligned}$$

Thus $x^{-1}C_G(a)x \subseteq C_G(b)$. On the other hand we have
$$a = xbx^{-1} = (x^{-1})^{-1}b(x^{-1})$$
and so, by the above argument,
$$xC_G(b)x^{-1} = (x^{-1})^{-1}C_G(b)(x^{-1}) \subseteq C_G(a).$$
Hence $C_G(b) \subseteq x^{-1}C_G(a)x$. $\qquad \square$

2.4 Homomorphisms
Problem 2.5 Let S and T be semigroups with laws of composition denoted by $*$ and $.$ respectively. Let $f: S \to T$ be a mapping such that
$$f(s_1 * s_2) = f(s_1) . f(s_2) \quad (s_1, s_2 \in S). \qquad (2.45)$$

Prove that $f(S)$, the image of S under f, is a semigroup under the law of composition on T.

Solution. Clearly $f(S) \neq \varnothing$. Let $x_1, x_2 \in f(S)$. Then, for some $y_1, y_2 \in S$, we have $x_1 = f(y_1)$, $x_2 = f(y_2)$. Thus

$$x_1 \cdot x_2 = f(y_1) \cdot f(y_2) = f(y_1 * y_2) \in f(S)$$

Thus $f(S)$ is closed under the associative law of composition in T and so $f(S)$ is a subsemigroup.

Such a mapping f is called a *homomorphism* (Greek *homo* − same, *morph* = form). $\qquad\square$

Problem 2.56 Let $\mathbb{R}$ be the set of real numbers considered as an additive group and let $\mathbb{R}^+$ be the set of strictly positive real numbers considered as a multiplicative group. Let $f : \mathbb{R}^+ \to \mathbb{R}$ be defined by $f(x) = \log x$ $(x \in \mathbb{R}^+)$. Prove that f is a homomorphism.

Solution. We make use of the well-known properties of the logarithm to obtain, for $x, y \in \mathbb{R}^+$,

$$f(xy) = \log xy = \log x + \log y$$
$$= f(x) + f(y). \qquad\square$$

In considering different groups that are possibly non-Abelian it is convenient to denote the formation of the product of two elements from any given group simply by juxtaposition of the elements, thereby in appearance, although not in reality, suppressing reference to the particular law of composition.

Problem 2.57 Let G and H be groups and let $f : G \to H$ be a homomorphism. Let G and H have identities e_G and e_H respectively. Prove that $f(e_G) = e_H$ and that $f(G)$ is a subgroup of H.

Solution. We have

$$f(e_G) = f(e_G e_G) = f(e_G) f(e_G)$$

and so $f(e_G) = e_H$. Further, for $x \in G$, we have

$$f(x) f(x^{-1}) = f(xx^{-1}) = f(e_G) = e_H$$

and thus

$$[f(x)]^{-1} = f(x^{-1})$$

where, on the left-hand side, the inverse is in H and on the right-hand side the inverse is in G. We know, from Problem 2.55, that $f(G)$ is a semigroup and that, since $e_H \in f(G)$, $f(G)$ is a monoid. Let $u \in f(G)$, then $u = f(v)$ $(v \in G)$ and so

$$u^{-1} = [f(v)]^{-1} = f(v^{-1}) \in f(G).$$

Hence $f(G)$ is a group.

We notice that a homomorphism 'preserves' multiplication and inverses. ☐

Problem 2.58 Let G and H be groups with identities e_G and e_H respectively. Let $f: G \rightarrow H$ be a homomorphism and let $K = \{x \in G : f(x) = e_H\}$. Prove that K is a normal subgroup of G.

Solution. Let $x, y \in K$, then

$$f(xy) = f(x)f(y) = e_H e_H = e_H$$
$$f(x^{-1}) = [f(x)]^{-1} = e_H^{-1} = e_H$$

and so K is a subgroup of G.

Let $x \in K$ and let $t \in G$, then

$$
\begin{aligned}
f(t^{-1}xt) &= f(t^{-1})f(xt) \\
&= [f(t)]^{-1}f(x)f(t) \\
&= [f(t)]^{-1}e_H f(t) \\
&= [f(t)]^{-1}f(t) \\
&= e_H
\end{aligned}
$$

and so $t^{-1}xt \in K$. Thus K is a normal subgroup of G.

The subgroup K is called the *kernel* of f. ☐

Problem 2.59 Let G and H be groups with identities e_G and e_H respectively. Let $f: G \rightarrow H$ be a homomorphism with kernel K. Let $x, y \in G$. Prove that $f(x) = f(y)$ if and only if $Kx = Ky$.

Solution. Suppose that $f(x) = f(y)$. Then $f(xy^{-1}) = f(x)f(y^{-1}) = f(x)[f(y)]^{-1} = f(x)[f(x)]^{-1} = e_H$. Thus $xy^{-1} \in K$ and hence, by Problem 2.46, $Kx = Ky$. Conversely if $Kx = Ky$ we have, by Problem 2.46, that $y = kx$ for some $k \in K$ and hence

$$f(y) = f(kx) = f(k)f(x) = e_H f(x) = f(x).$$

An important deduction may be made from the above. Let $a \in f(G) \subseteq H$. Then $a = f(z)$ for some $z \in G$. We assert that the subset of elements of G mapping onto a is precisely the coset $Kz = \{w : w = kz, k \in K\}$. This follows since $f(kz) = f(k)f(z) = e_H a = a$ and, on the other hand, $f(t) = a$ $(t \in G)$ implies that $Kt = Kz$ and so $t \in Kz$ (Problem 2.46).

Problem 2.60 Let $\mathbb{R}^*$, $\mathbb{C}^*$ be the multiplicative groups of nonzero real and complex numbers respectively. Prove that the mapping $f : \mathbb{C}^* \rightarrow \mathbb{R}^*$ given by $f(z) = |z|$ $(z \in \mathbb{C}^*)$ is a homomorphism and determine the kernel K of f.

Solution. By a standard result on the modulus of the product of complex numbers we have, for z_1, $z_2 \in \mathbb{C}^*$,

$$f(z_1 z_2) = |z_1 z_2| = |z_1||z_2| = f(z_1)f(z_2).$$

Thus f is a homomorphism and

$$K = \{z \in \mathbb{C}^* : f(z) = 1\}$$
$$= \{z \in \mathbb{C}^* : |z| = 1\}. \qquad \square$$

Problem 2.61 Let G be a group and suppose that the mapping $g : G \to G$ given by $g(x) = x^{-1}$ $(x \in G)$ is a homomorphism. Prove that G is Abelian.

Solution. Let $x, y \in G$. Then we have

$$xy = (x^{-1})^{-1}(y^{-1})^{-1} = (y^{-1}x^{-1})^{-1}$$
$$= g(y^{-1}x^{-1}) = g(y^{-1})g(x^{-1})$$
$$= (y^{-1})^{-1}(x^{-1})^{-1} = yx. \qquad \square$$

Problem 2.62 Let G and H be groups and let $f : G \to H$ be a homomorphism with kernel K. Let e_G be the identity of G. Prove that f is one–one if and only if $K = \{e_G\}$.

Solution. By definition f is one–one if and only if, for all $x, y \in G$, $f(x) = f(y)$ implies $x = y$.

Let us suppose first that f is one–one and let $x \in K$, we want to show that $x = e_G$. We have $Kx = K = Ke_G$ and so (Problem 2.59) $f(x) = f(e_G)$, since f is one–one we deduce that $x = e_G$.

Conversely suppose $K = \{e_G\}$ and let $a, b \in G$ be such that $f(a) = f(b)$, we want to show that $a = b$. We have (Problem 2.59) $Ka = Kb$ and so $\{a\} = \{e_G\}a = \{e_G\}b = \{b\}$. We conclude that $a = b$. $\qquad \square$

If the homomorphism $f : G \to H$ is one–one, i.e. if the kernel is trivial, we call f a *monomorphism* (Greek *mono* = alone). If the homomorphism $f : G \to H$ is such that f is onto, i.e. if $f(G) = H$, we call f an *epimorphism* (Greek *epi* = upon). A homomorphism f which is both one–one and onto is called an *isomorphism* (Greek *iso* = equal) and we say that G is isomorphic to H. The identity mapping ι_G of G onto G is an isomorphism. If T is a group of one element $T = \{e_T\}$ the mapping $h : G \to T$ given by $h(x) = e_T$ $(x \in G)$ is a homomorphism, such a homomorphism is *trivial*.

Problem 2.63 Let G be the subgroup of $\mathbb{C}$ generated by i and let H be the subgroup of $GL(2, \mathbb{R})$ generated by $\begin{pmatrix} 0 & 1 \\ -1 & 0 \end{pmatrix}$. Prove that there is an isomorphism between G and H.

42

Solution. Clearly we have

$$G = \{i, -1, -i, 1\}$$

$$H = \left\{ \begin{pmatrix} 0 & 1 \\ -1 & 0 \end{pmatrix}, \begin{pmatrix} -1 & 0 \\ 0 & -1 \end{pmatrix}, \begin{pmatrix} 0 & -1 \\ 1 & 0 \end{pmatrix}, \begin{pmatrix} 1 & 0 \\ 0 & 1 \end{pmatrix} \right\}.$$

The groups G and H are both cyclic of order 4 and we claim that the mapping g given by

$$g(i) = \begin{pmatrix} 0 & 1 \\ -1 & 0 \end{pmatrix},$$

$$g(-1) = \begin{pmatrix} -1 & 0 \\ 0 & -1 \end{pmatrix}, \qquad g(-i) = \begin{pmatrix} 0 & -1 \\ 1 & 0 \end{pmatrix},$$

$$g(1) = \begin{pmatrix} 1 & 0 \\ 0 & 1 \end{pmatrix}$$

is an isomorphism. The mapping is obviously one–one and onto and we may verify that the mapping is a homomorphism, e.g.

$$g[(-1)(-i)] = g(i) = \begin{pmatrix} 0 & 1 \\ -1 & 0 \end{pmatrix}$$

$$= \begin{pmatrix} -1 & 0 \\ 0 & -1 \end{pmatrix} \begin{pmatrix} 0 & -1 \\ 1 & 0 \end{pmatrix} = g(-1)g(-i). \qquad \square$$

Problem 2.64 Prove that the relation 'is isomorphic to' is an equivalence relation.

Solution. The relation is reflexive since the identity mapping on a group G maps G isomorphically onto G. To establish that the relation is symmetric, let G, H be groups such that G is isomorphic to H. We show that H is isomorphic to G. There exists a homomorphism $f : G \to H$ such that f is one–one and onto, and, by Problem 1.27, there exists an inverse mapping $g : H \to G$. We prove that g is a homomorphism. Let $y_1, y_2 \in H$, then there exist $x_1, x_2 \in G$ such that $f(x_1) = y_1$, $f(x_2) = y_2$ and, further, $g(y_1) = x_1, g(y_2) = x_2$. Hence $g(y_1 y_2) = g[f(x_1)f(x_2)] = g[f(x_1 x_2)] = (g \circ f)(x_1 x_2) = x_1 x_2 = g(y_1)g(y_2)$. Thus H is isomorphic to G. It is easy to verify that the relation is transitive and so, finally, we conclude that the relation is an equivalence relation. $\qquad \square$

Problem 2.65 Prove that there are two nonisomorphic groups of order 4 but that any group of order 4 must be isomorphic to one or other of these groups.

Solution. We know, from Problem 2.19(iv), that there is a cyclic group of order 4 and a noncyclic group, the Klein four-group. These groups are not isomorphic since, at the very least, isomorphic groups must have the same number of elements of any given order. We also know, from our construction of these groups, that any other group of order 4 is isomorphic to one or other of them. ◻

Problem 2.66 Let the group G have a normal subgroup H. Let G/H denote the set of cosets of H in G (since H is normal every left coset is a right coset and conversely). Define a law of composition $*$ on G/H by

$$Hx * Hy = Hxy \quad (x, y \in G). \tag{2.46}$$

Prove that, with this law of composition, G/H is a group.

Solution. It is not immediately obvious that our definition of a law of composition on G/H is meaningful as it seems to depend on a particular choice x, y of elements from the cosets Hx, Hy, or, equivalently, our definition is permissible only if we can infer from $Hu = Hx$ and $Hv = Hy$ $(u, v \in G)$ that $Huv = Hxy$; we proceed to establish this inference. We have $u = hx, v = ky$ for some $h, k \in H$ (Problem 2.46) and hence (Problem 2.21) we have

$$(uv)(xy)^{-1} = uvy^{-1}x^{-1} = hxkyy^{-1}x^{-1} = hxkx^{-1}.$$

But, since H is a normal subgroup of G, $xHx^{-1} = H$ and so $xkx^{-1} \in H$. Thus $(uv)(xy)^{-1} \in H$ and consequently $Huv = Hxy$ (Problem 2.46).

We now prove that G/H is a group.

G/H is certainly closed under $*$ and this law of composition is also associative for if $x, y, z \in G$ then

$$\begin{aligned}
(Hx * Hy) * Hz &= Hxy * Hz \\
&= H(xy)z \\
&= Hx(yz) \\
&= Hx * Hyz \\
&= Hx * (Hy * Hz).
\end{aligned}$$

The identity of G/H is the coset H for if e is the identity of G we have

$$H * Hx = He * Hx = Hex = Hx \quad (x \in G).$$

The inverse of $Hy\,(y \in G)$ is Hy^{-1} for

$$Hy^{-1} * Hy = Hy^{-1}y = He = H.$$

Thus, by Problem 2.14, G/H is a group. ◻

In the above we usually omit the $*$ and write $Hx * Hy = Hxy$, where of course, $Hx = xH$ and $Hy = yH$.

If $G = \bigcup_{\lambda \in \Lambda} Hx_\lambda$ is a coset decomposition, Λ being an indexing set, then

$$G/H = \{Hx_\lambda : \lambda \in \Lambda\} \tag{2.47}$$

and we have $|G/H| = |G : H|$.

Problem 2.67 Let G be a group and let H be a normal subgroup of G. Prove that the mapping $p : G \to G/H$ given by $p(x) = Hx$ $(x \in G)$ is an epimorphism.

Solution. The mapping p is clearly onto and we have, for $x, y \in G$, $p(xy) = Hxy = HxHy = p(x)p(y)$. $\qquad\qquad\qquad\qquad\qquad\square$

The group G/H is called a *factor-group* of G and the mapping p is sometimes called the *natural* homomorphism of G onto G/H. To every normal subgroup H of G there corresponds such an epimorphism and conversely.

Problem 2.68 The following array gives the law of multiplication for a group G of order 8 called the *quaternion group*. Find the centre $Z(G)$ of G and show that $G/Z(G)$ is isomorphic to the Klein four-group.

	e	a	b	c	d	f	g	h
e	e	a	b	c	d	f	g	h
a	a	e	c	b	f	d	h	g
b	b	c	a	e	g	h	f	d
c	c	b	e	a	h	g	d	f
d	d	f	h	g	a	e	b	c
f	f	d	g	h	e	a	c	b
g	g	h	d	f	c	b	a	e
h	h	g	f	d	b	c	e	a

$$(2.48)$$

Solution. We may determine $Z(G)$ by inspection for, if $x \in Z(G)$ then pre- or post-multiplication by x of any other element yields the same result or, equivalently, $x \in Z(G)$ if and only if the row of the array through x is the transpose of the column through x. Hence we see that $Z(G) = \{e, a\}$.

We first obtain a coset decomposition of $Z(G)$ in G and, for later convenience, we denote the cosets by Greek letters, thus we have

$$\varepsilon = \{e, a\} = Z(G)$$
$$\alpha = \{e, a\}b = \{b, c\}$$
$$\beta = \{e, a\}d = \{d, f\}$$
$$\gamma = \{e, a\}g = \{g, h\}.$$

Then we have, for example,

$$\alpha\beta = [\{e,a\}b][\{e,a\}d]$$
$$= \{e,a\}bd$$
$$= \{e,a\}g$$
$$= \gamma$$
$$\alpha\gamma = [\{e,a\}b][\{e,a\}g]$$
$$= \{e,a\}bg$$
$$= \{e,a\}f$$
$$= \{f,d\}$$
$$= \beta$$

and continuing in this way we obtain the array

	ε	α	β	γ
ε	ε	α	β	γ
α	α	ε	γ	β
β	β	γ	ε	α
γ	γ	β	α	ε

We know that this array represents the Klein four-group. ☐

Problem 2.69 Let G be a group and let H be a subgroup of G contained in the centre of G. If G/H is cyclic prove that G is Abelian.

Solution. We observe first that H *is* normal in G. Since G/H is cyclic every element of G/H is a power of some fixed element, i.e. there exists a coset Hc ($c \in G$) such that any other coset is of the form $(Hc)^n$. But

$$(Hc)^n = HcHc\ldots Hc$$
$$= Hc^n$$

and thus

$$G = \bigcup_{n \in \mathbb{Z}} Hc^n.$$

Now let $x, y \in G$. Then there exist $h, k \in H$, $r, s \in \mathbb{Z}$ such that $x = hc^r$,

$y = kc^s$. Hence

$$xy = (hc^r)(kc^s)$$
$$= h(c^r k)c^s$$
$$= h(kc^r)c^s$$
$$= (hk)(c^r c^s)$$
$$= (kh)(c^s c^r)$$
$$= k(hc^s)c^r$$
$$= k(c^s h)c^r$$
$$= (kc^s)(hc^r)$$
$$= yx.$$

We could have used this result in Problem 2.68 for there we had a non-Abelian group G such that $G/Z(G)$ had order 4. There are, up to isomorphism, two groups of order 4, the Klein four-group and cyclic group the latter of which could not have arisen by this problem. □

Problem 2.70 Let G and H be groups and let $f : G \to H$ be an epimorphism having kernel K. Prove that H and G/K are isomorphic.

Solution. We seek to define a mapping $g : G/K \to H$ in such a way that g is an isomorphism.

We observe first that if $x, y \in G$ and $Kx = Ky$ then $f(x) = f(y)$ for we have $x = ky$ ($k \in K$) and so, as $f(k)$ is the identity of H, $f(x) = f(ky) = f(k)f(y) = f(y)$. Thus we see that $f(x)$ does not depend on x but on the coset Kx to which x belongs. Consequently we define, unambiguously, the mapping g by

$$g(Kx) = f(x) \quad (x \in G).$$

We require now to verify that g has the required properties:

1 g *is onto.* Let $h \in H$. Then, since f is onto there exists $z \in G$ such $f(z) = h$ and so $g(Kz) = f(z) = h$.
2 g *is one–one.* Let $a, b \in G$ and suppose $g(Ka) = g(Kb)$. Then since $f(a) = f(b)$, $f(a^{-1}b)$ is the identity of H and so $a^{-1}b \in K$ which implies that $Ka = Kb$.
3 g *is a homomorphism.* Let $u, v \in G$.
   Then
   $$g(KuKv) = g(Kuv)$$
   $$= f(uv)$$
   $$= f(u)f(v)$$
   $$= g(Ku)g(Kv).$$

This result is known as the *first isomorphism theorem*. □

Problem 2.71   Let G and H be groups and let $f : G \to H$ be a homomorphism having kernel K. Prove that $f(G)$ and G/K are isomorphic.

Solution. We have $f : G \to f(G)$ as an epimorphism. We now apply the previous result. □

Problem 2.72   Let G be a finite group and let H be a group onto which G can be mapped by an epimorphism. Prove that $|H|$ divides $|G|$.

Solution. Letting K be the kernel of the epimorphism we have that G/K is isomorphic to H and so

$$|G| = |K||G : K| = |K||G/K| = |K||H|.$$ □

A nontrivial group G is said to be *simple* if G has no proper normal subgroups, equivalently any epimorphism of G onto a nontrivial group is necessarily an isomorphism. If G is Abelian and simple then G has no proper subgroups and so $|G|$ is a prime.

Problem 2.73   Let G be a group and let N, H be subgroups of G, N being normal in G. Prove that $N \cap H$ is a normal subgroup of H.

Solution. $N \cap H$ is certainly a subgroup of H. Let $a \in N \cap H$ and let $x \in H$. Then $x^{-1}ax \in H$ and, since N is normal in G, $x^{-1}ax \in N$. Hence $x^{-1}ax \in N \cap H$ and so $N \cap H$ is normal in H. □

Problem 2.74   Let G be a group and let N, H be subgroups of G, N being normal in G. Prove that NH/N and $H/(N \cap H)$ are isomorphic.

Solution. We note first that NH is a subgroup of G and that the factorgroups NH/N and $H/(N \cap H)$ are defined.

We now try to define a homomorphism from H to NH/N in such a way that the homomorphism is surjective and has $N \cap H$ as kernel, we then apply the first isomorphism theorem. Let $p : H \to NH/N$ be defined by $p(h) = Nh \ (h \in H)$. We claim that p is an epimorphism. First p is a homomorphism since, for $h_1, h_2 \in H$, $p(h_1 h_2) = Nh_1 h_2 = Nh_1 Nh_2 = p(h_1)p(h_2)$. Second, p is surjective since if $x \in NH/N$ we have $x = Nu$ where $u \in NH$ and so $u = nk \ (n \in N, k \in H)$ giving $p(k) = Nk = Nnk = Nu$. Let K be the kernel of p. Then K is a normal subgroup of H and, in fact, $K = \{y : y \in H, \ p(y) = N\} = \{y : y \in H, \ Ny = N\} = \{y : y \in H, \ y \in N\} = H \cap N$. By the first isomorphism theorem $H/(N \cap H)$ and NH/H are isomorphic.

This result is known as the *second isomorphism theorem*. □

48

1. Let S be a nonempty set in which a multiplication $*$ is defined by $s * t = s \, (s, t \in S)$. Prove that S is a semigroup. If S is a monoid prove that S consists of precisely one element.

2. Let S be the set of 2×2 matrices over $\mathbb{Q}$. Let $\mathbf{m}$ be a given matrix and introduce a multiplication $*$ in S by defining $\mathbf{s} * \mathbf{t} = \mathbf{smt} \, (\mathbf{s}, \mathbf{t} \in S)$ where on the right-hand side ordinary matrix multiplication is to be understood. Prove that S is a semigroup and that, if $\mathbf{m}$ is nonsingular, S is a monoid.

3. Let S be a finite semigroup and let $s \in S$. Prove that there exists m, $n \in \mathbb{Z}, 0 < 2m < n$, such that $s^m = s^n$. If $t = s^{n-m}$ prove that $t^2 = t$.

4. Let G be a group and let $x, y, z \in G$. Prove that
$$(yx^{-1})^{-1}(yx)^2(zyx)^{-1}(xz^{-1})^{-1} = x = xy^{-1}x^{-1}yx^{-1}zx(x^{-1}zx)^{-1}y^{-1}xy.$$

5. Prove that the set G of 2×2 matrices of the form
$$\begin{pmatrix} -c+ib & c+id \\ -c+id & a-ib \end{pmatrix}$$
where $a, b, c, d \in \mathbb{R}, a^2 + b^2 + c^2 + d^2 \neq 0$, is a group under matrix multiplication.

6. Let H be a nonempty subset of the group G. Prove that H is a subgroup if and only if $xy^{-1} \in H$ for all $x, y \in H$.

7. Prove that the group of Problem 2.6 has five subgroups.

8. Prove that the group of Problem 2.27 has six subgroups.

9. Let H be a subgroup of the group G. Let $N_G(H) = \{x : x \in G, x^{-1}Hx = H\}$ and $C_G(H) = \{x : x \in G, x^{-1}hx = h$ for all $h \in H\}$. Prove that $C_G(H)$ is a normal subgroup of the subgroup $N_G(H)$. If $|G : N_G(H)|$ is finite and equals n prove that H has precisely n conjugates (cf. Problem 2.53).

10. For the quaternion group G given in Problem 2.68 prove that c generates a subgroup H of order 4 and find a coset decomposition of H in G.

11. Let G be a cyclic group of order 6 generated by x. Let H, K be the subgroups generated by x^2, x^3 respectively. Prove that $|H| = 3, |K| = 2$, $G = HK$, and that $H \cap K$ is trivial.

12. Let $GL(2, \mathbb{R})$ be the group of 2×2 nonsingular matrices over $\mathbb{R}$.

Prove that the matrices of the form

$$\begin{pmatrix} a & x \\ 0 & b \end{pmatrix} \quad (ab \neq 0)$$

form a subgroup H. Prove that the centre of H is the centre of $GL(2, \mathbb{R})$ and that H is its own normaliser.

13. Let N_1, N_2 be normal subgroups of the group G. Prove that $N_1 N_2$ is a normal subgroup of G. If $N_1 \cap N_2$ is trivial prove that $x_1 x_2 = x_2 x_1$ for all $x_1 \in N_1$, $x_2 \in N_2$.

14. Let H, K be subgroups of the group G such that $H \subseteq K \subseteq G$. Prove that

$$|G : H| = |G : K||K : H|.$$

15. Let H be a normal subgroup of the group G of index 4. Prove that G/H is Abelian.

16. Let a be an element of the group G and let $T = \{x : x \in G, x^{-1}ax = a \text{ or } x^{-1}ax = a^{-1}\}$. Prove that T is a subgroup of G and that if $C_G(a)$ is the centraliser of a in G then $|T : C_G(a)| = 1$ or 2.

17. Let G, H, K be groups and let f, g be homomorphisms such that $G \xrightarrow{f} H, H \xrightarrow{g} K$. Prove that $g \circ f$ is a homomorphism.

18. Suppose that the mapping f on the group G given by $f(x) = x^2 (x \in G)$ is a homomorphism. Prove that G is Abelian.

19. Let N be a normal subgroup of the group G. Let $x \in G$. If x has finite order in G prove that Nx has finite order in G/N. By considering $\mathbb{Z}$ show that the converse implication is false.

20. Let $\mathbb{R}$ be the set of real numbers considered as an additive group and let $\mathbb{R}^+$ be the set of strictly positive real numbers considered as a multiplicative group. Let $g : \mathbb{R} \to \mathbb{R}^+$ be defined by $g(x) = e^x$ ($x \in \mathbb{R}$). Prove that g is an isomorphism. (cf. Problem 2.56.)

21. Let H, N be subgroups of the group G, N being normal in G. Prove that

$$|HN||H \cap N| = |H||N|.$$

22. Let $Z(G)$ be the centre of the group G. Let a be an element of G such that $x^{-1}a^{-1}xa \in Z(G)$ for all $x \in G$. Prove that the mapping f given by $f(x) = x^{-1}a^{-1}xa$ is a homomorphism of G into $Z(G)$. If $a \notin Z(G)$ prove that f is nontrivial.

23. Let H be a proper subgroup of the nontrivial finite group G. Let H have n conjugates $H = H_1, H_2, \ldots, H_n$. Prove that the set $H_1 \cup H_2 \cup \ldots \cup H_n$ contains, at most, $n(|H|-1)+1$ elements of G. Deduce that $G \neq \bigcup_{i=1}^{n} H_i$.

Chapter 3

Direct Products, Abelian Groups

3.1 Direct products This section is concerned with a simple method by which groups may be constructed from given groups.

Problem 3.1 Let G, H be groups. A law of composition is defined on $G \times H$ by

$$(g_1, h_1)(g_2, h_2) = (g_1 g_2, h_1 h_2) \quad (g_i \in G, \quad h_i \in H; \quad i = 1, 2). \quad (3.1)$$

Prove that, under this law of composition, $G \times H$ is a group.

Solution. By definition $G \times H$ is closed under the law of composition. To establish associativity we let $g_j \in G$, $h_j \in H$ ($j = 1, 2, 3$) and then

$$[(g_1, h_1)(g_2, h_2)](g_3, h_3) = (g_1 g_2, h_1 h_2)(g_3, h_3)$$
$$= ((g_1 g_2)g_3, (h_1 h_2)h_3) = (g_1 (g_2 g_3), h_1 (h_2 h_3))$$
$$= (g_1, h_1)(g_2 h_3, h_2 h_3) = (g_1, h_1)[(g_2, h_2)(g_3, h_3)]. \quad (3.2)$$

The identity element of $G \times H$ is (e_G, e_H) where e_G, e_H are the identity elements of G, H respectively. The inverse of (g, h) $(g \in G, h \in H)$ is (g^{-1}, h^{-1}). Thus $G \times H$ is a group.

We call $G \times H$ with this 'component-wise' multiplication the *direct product of G and H*. ∎

Problem 3.2 Let G, H be groups. Prove that $G \times H$ and $H \times G$ are isomorphic. Show that $G \times H$ is Abelian if and only if both G, H are Abelian.

Solution. It is an easy verification that $G \times H$ and $H \times G$ are isomorphic under the mapping $(g, h) \to (h, g)$.

Let (g_1, h_1), (g_2, h_2) be elements of $G \times H$. Then $(g_1, h_1)(g_2, h_2) = (g_1 g_2, h_1 h_2)$ and so $(g_1, h_1)(g_2, h_2) = (g_2 g_1, h_2 h_1)$ if and only if $g_1 g_2 = g_2 g_1$ and $h_1 h_2 = h_2 h_1$ for all $g_1, g_2 \in G$, $h_1, h_2 \in H$. Thus $G \times H$ is Abelian if and only if G, H are Abelian. ∎

Problem 3.3 Let $G \times H$ be the direct product of the groups G, H. Let $\bar{G} = \{(g, e_H) : g \in G\}$, $\bar{H} = \{(e_G, h) : h \in H\}$. Prove that $\bar{G}$ is isomorphic to G and that $\bar{H}$ is isomorphic to H.

Solution. The mapping $f : G \to \bar{G}$ given by $f(g) = (g, e_H)$ is clearly bijective and f is also a homomorphism for if $g_1, g_2 \in G$ we have

$$f(g_1 g_2) = (g_1 g_2, e_H) = (g_1, e_H)(g_2, e_H) = f(g_1) f(g_2).$$

Thus G and $\bar{G}$ are isomorphic and similarly so are H and $\bar{H}$.

We remark that $\bar{G}, \bar{H}$ are normal subgroups of $G \times H$ such that $G \times H = \bar{G}\bar{H}$ and $\bar{G} \cap \bar{H}$ is trivial. We also have that $(G \times H)/\bar{G}$ is isomorphic to H and $(G \times H)/\bar{H}$ is isomorphic to G. $\qquad\square$

Problem 3.4 Let X, Y be cyclic groups of order 2. Prove that $X \times Y$ is isomorphic to the Klein four-group.

Solution. Let X, Y have generators x, y respectively. Then
$$X \times Y = \{(e_X, e_Y), (e_X, y), (x, e_Y), (x, y)\}.$$
The four-group G is given by the array (2.28). We may then verify that the mapping $f : G \to X \times Y$, given by
$$f(e) = (e_X, e_Y), \qquad f(a) = (e_X, y), \qquad f(b) = (x, e_Y), \qquad f(c) = (x, y)$$
is an isomorphism. $\qquad\square$

Problem 3.5 Let X, Y be cyclic groups of orders 2, 3 respectively. Find the orders of the elements of $X \times Y$.

Solution. Let X, Y have generators x, y respectively. Then (e_X, e_Y) has order 1. (x, e_Y) has order 2. $(e_X, y), (e_X, y^2)$ have order 3. $(x, y), (x, y^2)$ have order 6.

We observe that $X \times Y$ is cyclic. In general if X, Y are finite cyclic groups of relatively coprime orders then $X \times Y$ is also cyclic. $\qquad\square$

From two groups G, H we have composed a third group $G \times H$. We also wish to know when a given group G can be decomposed into a direct product.

Problem 3.6 Let H_1, H_2 be normal subgroups of a group G such that $G = H_1 H_2$ and $H_1 \cap H_2 = \{e\}$ where e is the identity of G. Prove that G is isomorphic to the direct product $H_1 \times H_2$.

Solution. We observe first that the elements of H_1 commute with the elements of H_2. To see this let $x_1 \in H$, $x_2 \in H$ and then
$$x_1^{-1}(x_2^{-1} x_1 x_2) = (x_1^{-1} x_2^{-1} x_1)x_2 \in H_1 \cap H_2$$
from which we deduce that $x_1^{-1}x_2^{-1}x_1 x_2 = e$ or, equivalently, that x_1, x_2 commute.

We now define $f : H_1 \times H_2 \to G$ by $f(x_1, x_2) = x_1 x_2 (x_1 \in H_1, x_2 \in H_2)$. Then f is a homomorphism since we have, for $y_1 \in H_1$, $y_2 \in H_2$,
$$f[(x_1, x_2)(y_1, y_2)] = f[(x_1 y_1, x_2 y_2)]$$
$$= x_1 y_1 x_2 y_2 = x_1 x_2 y_1 y_2 = f(x_1, x_2)f(y_1, y_2).$$
f is surjective for, if $g \in G = H_1 H_2$ then $g = g_1 g_2$ $(g_1 \in H, g_2 \in H_2)$ and then $f(g_1, g_2) = g_1 g_2 = g$. f is injective for if $f(h_1, h_2) = e$ $(h_1 \in H_1, h_2 \in H_2)$ then $h_1 h_2 = e$ and so $h_1 = h_2^{-1} \in H_1 \cap H_2$ from which we deduce that $h_1 = h_2 = e$.

If G has subgroups H_1, H_2 with the above properties then we often say briefly that G is the direct product of its subgroups H_1, H_2. $\square$

For convenience we shall denote the identity of any group by e.

Problem 3.7 Let G be an Abelian group of order 9. Prove that G is either cyclic or is isomorphic to a direct product of two groups each of order 3.

Solution. If G has an element of order 9 then G is cyclic. Suppose G has no element of order 9 and let $x \in G$, $x \neq e$. The order of x divides 9 and so x has order 3. Let H be the subgroup generated by x, $H = \{e, x, x^2\}$. Let $y \in G$, $y \notin H$ and let K be the subgroup generated by y, $K = \{e, y, y^2\}$. Then $H \cap K = \{e\}$ for $y \notin H$ and if $y^2 \in H$ we would have $y = y^3 y = (y^2)^2 \in H$ which is false. Hence we have (Exercise 21 in Chapter 2)

$$|HK| = |HK||H \cap K| = |H||K| = 3 \times 3 = 9.$$

Thus $G = HK$ and so G is the direct product of H and K.

We remark that a group of order p^2 where p is a prime its necessarily Abelian. $\square$

The notion of the direct product of two groups may be extended to the direct product of a finite number of groups. Thus if $G_1, G_1, \ldots, G_n$ are n groups we form the Cartesian product $G_1 \times G_2 \times \ldots \times G_n$ and define a law of composition by

$$(g_1, g_2, \ldots, g_n)(g_1', g_2', \ldots, g_n') = (g_1 g_1', g_2 g_2', \ldots, g_n g_n')$$

$$(g_i, g_i' \in G_i \quad i = 1, 2, \ldots, n). \tag{3.3}$$

The group laws are easily verified and we speak of the *direct product* $G_1 \times G_2 \times \ldots \times G_n$, the law of composition being as above. Let $\bar{G}_i$ be the subset of $G_1 \times G_2 \times \ldots \times G_n$ having the elements of G_i at the ith component and identity elements elsewhere, then G_i and $\bar{G}_i$ are isomorphic, $G_1 \times G_2 \times \ldots \times G_n = \bar{G}_1 \bar{G}_2 \ldots \bar{G}_n$ and $\bar{G}_i \cap (\bar{G}_1 \bar{G}_2 \ldots \bar{G}_{i-1} \bar{G}_{i+1} \ldots G_n) = \{e\} (i = 1, 2, \ldots, n)$; the subgroups $\bar{G}_1, \bar{G}_2, \ldots, \bar{G}_n$ generate $G_1 \times G_2 \times \ldots \times G_n$ and have the property that the subgroup generated by any $n-1$ of them intersects the remaining subgroup trivially. We notice also that $(G_1 \times G_2 \times \ldots \times G_n)/\bar{G}_i$ is isomorphic to $G_1 \times G_2 \times \ldots \times G_{i-1} \times G_{i+1} \times \ldots \times G_n$ $(i = 1, 2, \ldots, n)$.

Conversely we may show that if the group G has normal subgroups $H_1, H_2, \ldots, H_n$ such that $G = H_1 H_2 \ldots H_n$ and $H_i \cap H_1 H_2 \ldots H_{i-1} H_{i+1} \ldots H_n = \{e\}$ $(i = 1, 2, \ldots, n)$ then G is isomorphic to the direct product $H_1 \times H_2 \times \ldots \times H_n$, briefly we say G is the direct product of its subgroups $H_1, H_2, \ldots, H_n$. It is clear that if G is the direct product of its subgroups $H_1, H_2, \ldots, H_n$ then certainly $H_i \cap H_j = \{e\}$ $(i \neq j)$.

Problem 3.8 Give an example of a group G having three normal subgroups H_1, H_2, H_3 such that $G = H_1 H_2 H_3$ and $H_i \cap H_j = \{e\}$ ($i \neq j$) but nevertheless G is not the direct product $H_1 \times H_2 \times H_3$.

Solution. Let G be the Klein four-group and let H_1, H_2, H_3 be the three subgroups each of order 2, G is evidently not the direct product of these groups. $\square$

An alternative and useful characterisation is that G is the direct product of its normal subgroups H_1, H_2,..., H_n if and only if for all $x \in G$ there exist *unique* elements $x_i \in H_i$ ($i = 1, 2, ..., n$) such that $x_i x_j = x_j x_i$ ($i \neq j$) and $x = x_1 x_2 ... x_n$.

3.2 Abelian groups
A group G is Abelian if any two elements of G commute. In an Abelian group G we have for $x, y \in G$

$$(xy)^n = x^n y^n \quad (n = 0, \pm 1, ...). \tag{3.4}$$

Problem 3.9 Let G be an Abelian group. Prove that $f : G \to G$ given by $f(x) = x^2$ ($x \in G$) is a homomorphism and that if G is a finite group of odd order f is an isomorphism.

Solution. $f(xy) = (xy)^2 = x^2 y^2 = f(x)f(y)$ ($x, y \in G$) and so f *is a* homomorphism. If G has odd order d we write $d = 2c+1$ ($c \in \mathbb{Z}$) and so, if $u \in G$,

$$u = u^d u = u^{d+1} = (u^{c+1})^2 = f(u^{c+1})$$

thus showing that f is surjective. We may now assert that f is injective, for a surjective mapping of a finite set onto itself is necessarily injective; alternatively, if $t \in G$ belongs to the kernel of f then $t^2 = e$ and so, as G has odd order, $t = e$. $\square$

Problem 3.10 Let G be an Abelian group and let n be a positive integer. Prove that $\{x : x \in G, x^n = e\}$ is a subgroup of G.

Solution. Let $x, y \in G$ be such that $x^n = y^n = e$. Then $(xy)^n = x^n y^n = ee = e$, $(x^{-1})^n = (x^n)^{-1} = e$. Thus the subset of G consisting of elements of orders dividing n is a subgroup of G. $\square$

Problem 3.11 Let G be an Abelian group and let n be a positive integer. Prove that $\{z : z = x^n, x \in G\}$ is a subgroup of G.

Solution. Since we have, for all $x, y \in G$, $x^n y^n = (xy)^n$, $(x^n)^{-1} = (x^{-1})^n$ it follows that the subset of G consisting of the nth powers of the elements of G is a subgroup of G. $\square$

Problem 3.12 Let T be the subset of the Abelian group G consisting of all elements of finite order. Prove that T is a subgroup of G and that G/T has no nontrivial elements of finite order.

Solution. Obviously $T \neq \varnothing$ since $e \in T$. Let $x, y \in T$. Then there exist $m, n \geqslant 1$ such that $x^m = y^n = e$ and so $(xy)^{mn} = x^{mn} y^{mn} = (x^m)^n (y^n)^m = ee = e$. Thus T is a subgroup.

We want now to show that if zT $(z \in G)$ has finite order then $zT = T$ or, equivalently, $z \in T$. Suppose $(zT)^d = T$ $(d \geqslant 1)$. Then $z^d T = (zT)^d = T$ and so $z^d = t \in T$. By assumption there exists c $(c \geqslant 1)$ such that $t^c = e$ and this implies that $z^{cd} = (z^d)^c = t^c = e$ from which we deduce that $z \in T$.

□

In any group, whether Abelian or not, an element of finite order is often said to be a *torsion* or *periodic* element; otherwise the element is said to be *torsion-free*. If all elements of a group are periodic, the group is called *periodic*; if there are no nontrivial elements of finite order the group is said to be *torsion-free*. Any finite group is periodic, the infinite cyclic group is torsion-free. In an Abelian group G the torsion elements form a subgroup T called the *torsion* subgroup.

Problem 3.13 Let G be the infinite cyclic group generated by x. Let $Y = \{y : y = x^{12a + 30b + 42c}; a, b, c \in \mathbb{Z}\}$. Prove that Y is the cyclic subgroup of G generated by x^6.

Solution. Let H be the subgroup of G generated by x^6. Then $x^{12a + 30b + 42c} = (x^6)^{2a + 5b + 7c} \in H$ and so $Y \subseteq H$. On the other hand, by the usual Euclidean algorithm for the greatest common divisor, there exist a_0, b_0, c_0 such that

$$12a_0 + 30b_0 + 42c_0 = 6$$

for example $a_0 = -2, b_0 = 1, c_0 = 0$ or $a_0 = 2, b_0 = 5, c_0 = -4$. Thus $x^6 = x^{12a_0 + 30b_0 + 42c_0} \in Y$ and so $H \subseteq Y$. Hence $Y = H$. □

Problem 3.14 Let $x_1, x_2, \ldots, x_n$ be elements of the Abelian group G. Let $H = \{z : z = x_1^{c_1} x_2^{c_2} \ldots x_n^{c_n}, c_i \in \mathbb{Z}; i = 1, 2, \ldots, n\}$. Prove that H is a subgroup of G containing $\{x_1, x_2, \ldots, x_n\}$.

Solution. The first assertion follows immediately from the commutativity of G since

$$(x_1^{c_1} x_2^{c_2} \ldots x_n^{c_n})(x_1^{d_1} x_2^{d_2} \ldots x_n^{d_n}) = x_1^{c_1 + d_1} x_2^{c_2 + d_2} \ldots x_n^{c_n + d_n}$$

and

$$(x_1^{c_1} x_2^{c_2} \ldots x_n^{c_n})^{-1} = x_1^{-c_1} x_2^{-c_2} \ldots x_n^{-c_n}.$$

The rest is obvious.

We observe that any subgroup of G containing $\{x_1, x_2, \ldots, x_n\}$ necessarily contains H; H is therefore the least such subgroup. The elements $x_1, x_2, \ldots, x_n$ are said to *generate* H and H is said to be *finitely generated*; if $n = 1$ we recover our previous notion of a generator of a cyclic group.

If $H = G$ then G is said to be finitely generated, thus G is finitely generated if and only if G has a subset $\{x_1, x_2, \ldots, x_n\}$ such that every element of G is expressible as $x_1^{a_1} x_2^{a_2} \ldots x_n^{a_n}$ for suitable $a_1, a_2, \ldots, a_n \in \mathbb{Z}$. Any finite Abelian group is finitely generated. $\qquad \square$

An Abelian group G is said to have a *finite basis* if there exist $x_i \in G$ ($i = 1, 2, \ldots, n$) such that $x_1, x_2, \ldots, x_n$ generate G and such that $x_1^{a_1} x_2^{a_2} \ldots x_n^{a_n} = e$ implies $x_i^{a_i} = e$ $(i = 1, 2, \ldots, n)$.

Problem 3.15 Let G be the direct product of two cyclic subgroups. Prove that G has a basis of two elements.

Solution. By assumption G has cyclic normal subgroups H_1, H_2 such that $G = H_1 H_2$ and $H_1 \cap H_2 = \{e\}$. Let x_i generate H_i $(i = 1, 2)$. Then
$$G = \{z : z = x_1^{a_1} x_2^{a_2}; \quad a_1, a_2 \in \mathbb{Z}\}$$
and if $x_1^{a_1} x_2^{a_2} = e$ we have $x_1^{a_1} = x_2^{-a_2} \in H_1 \cap H_2$ and so $x_1^{a_1} = x_2^{a_2} = e$.

This result obviously extends to the direct product of any finite number of cyclic groups. $\qquad \square$

Problem 3.16 Let G be the direct product of the infinite cyclic subgroups generated by a, b respectively. Let $c = a^4 b^5$, $d = a^2 b^3$ and let $H =$ the subgroup generated by c, d. Prove that H is a proper subgroup of G having $\{c, d\}$ as a basis.

Solution. Suppose we have $c^m d^n = e$ for some m, n. Then
$$a^{4m + 2n} b^{5m + 3n} = (a^4 b^5)^m (a^2 b^3)^n = c^m d^n = e$$
and so, as $\{a, b\}$ is a basis of G, $a^{4m + 2n} = b^{5m + 3n} = e$. But a, b generate infinite cyclic groups and so we must have $4m + 2n = 0$, $5m + 3n = 0$ from which we deduce that $m = n = 0$ and hence $c^m = d^n = e$. Thus $\{a, b\}$ is a basis of H.

If $H = G$ then, in particular, $a \in H$ and so, for some r, s, we would have
$$a = (a^4 b^5)^r (a^2 b^3)^s = a^{4r + 2s} b^{5r + 3s}$$
which implies that
$$4r + 2s = 1, \qquad 5r + 3s = 0.$$
These equations have no *integral* solutions and so we have a contradiction. Thus $H \neq G$.

We observe that H is the direct product of the infinite cyclic subgroups generated by c, d. $\qquad \square$

A group G is said to be a *finitely generated free Abelian group* if G is the direct product of a finite number of infinite cyclic subgroups. Equivalently a finitely generated free Abelian group G has a finite basis $\{x_1, x_2, \ldots, x_n\}$ such that $x_i^{a_i} = e$ implies $a_i = 0$ $(i = 1, 2, \ldots, n)$, such a basis of G is called a *free basis*; a free Abelian group is certainly torsion-free.

Problem 3.17 Let G be the direct product of three infinite cyclic subgroups A, B, C generated by a, b, c respectively. Let M, N, P be the cyclic groups generated by a^2b^3c, $ab^{-4}c^2$, $a^{-1}c^{-1}$ respectively. Prove that M, N, P are infinite cyclic groups such that $G = M \times N \times P$.

Solution. We remark first that, as $G = A \times B \times C$, every element of G is uniquely expressible as $a^u b^v c^w$ for suitable u, v, w ($u, v, w \in \mathbb{Z}$). To show that $G = M \times N \times P$ we have to show that every such element of G is expressible as

$$(a^2b^3c)^x (ab^{-4}c^2)^y (a^{-1}c^{-1})^z = a^{2x+y-z} b^{3x-4y} c^{x+2y-z}$$

for suitable $x, y, z \in \mathbb{Z}$ (i.e. that $G = MNP$) and, furthermore, that this expression is unique (i.e. that $M \cap NP = N \cap MP = P \cap MN = \{e\}$). However every element of G may be so expressed for the linear equations

$$2x + y - z = u$$
$$3x - 4y = v \qquad\qquad (3.5)$$
$$x + 2y - z = w$$

admit a unique solution with integral coefficients, namely

$$x = 4u - v - 4w$$
$$y = 3u - v - 3w \qquad\qquad (3.6)$$
$$z = 10u - 3v - 11w.$$

Hence $G = M \times N \times P$.

It may be shown that if G is a finitely generated free Abelian group with free basis $\{x_1, x_2, \ldots, x_n\}$ and if $\{y_1, y_2, \ldots, y_m\}$ is a subset of G where

$$y_i = x_1^{a_{i_1}} x_2^{a_{i_2}} \ldots x_n^{a_{i_n}} \quad (i = 1, 2, \ldots, m)$$

then $\{y_1, y_2, \ldots, y_m\}$ is also a free basis of G if and only if $m = n$ and the matrix (a_{ij}) of the indices has determinant equal to ± 1. Thus the number of elements in any free basis is independent of the basis chosen; this number we call the *rank* of G; a free Abelian group of rank n is isomorphic to the direct product of n infinite cyclic groups and two free Abelian groups are isomorphic if and only if they have the same rank.

The rank of a finitely generated free Abelian group is analogous to the dimension of a finite-dimensional vector space over $\mathbb{R}$. Notice however that an infinite cyclic group *and all its proper subgroups* have the same rank, 1, whereas proper subspaces of a vector space necessarily have lower dimension than the whole space. $\qquad\qquad \square$

An important and characteristic property of free Abelian groups is given in the following problem.

Problem 3.18 Let F be a free Abelian group of rank n having $X = \{x_1,$

58

$x_2, \ldots, x_n\}$ as a free basis. Let G be an Abelian group and let $f : X \to G$ be a mapping. Prove that f 'extends' to a homomorphism $g : F \to G$.

Solution. What we are trying to do is to define a homomorphism g on F that coincides with the mapping f on X, g is then said to be an *extension* of f. Thus we want to have $g(x_i) = f(x_i)$ $(i = 1, 2, \ldots, n)$ and if g is to be a homomorphism, of necessity, we must define

$$g(x_1^{a_1} x_2^{a_2} \ldots x_n^{a_n}) = [g(x_1)]^{a_1} [g(x_2)]^{a_2} \ldots [g(x_n)]^{a_n}$$
$$= [f(x_1)]^{a_1} [f(x_2)]^{a_2} \ldots [f(x_n)]^{a_n}.$$

Now as every element of F is uniquely representable in the form $x_1^{a_1} x_2^{a_2} \ldots x_n^{a_n}$, g is indeed well-defined as a mapping. But g is also a homomorphism since

$$g[(x_1^{a_1} x_2^{a_2} \ldots x_n^{a_n})(x_1^{b_1} x_2^{b_2} \ldots x_n^{b_n})]$$
$$= g(x_1^{a_1+b_1} x_2^{a_2+b_2} \ldots x_n^{a_n+b_n})$$
$$= [f(x_1)]^{a_1+b_1} [f(x_2)]^{a_2+b_2} \ldots [f(x_n)]^{a_n+b_n}$$
$$= ([f(x_1)]^{a_1} [f(x_2)]^{a_2} \ldots [f(x_n)]^{a_n})([f(x_1)]^{b_1} [f(x_2)]^{b_2} \ldots [f(x_n)]^{b_n})$$
$$= g(x_1^{a_1} x_2^{a_2} \ldots x_n^{a_n}) \, g(x_1^{b_1} x_2^{b_2} \ldots x_n^{b_n}). \tag{3.8}$$

Incidentally we see why F is said to be 'free' and to have a 'free' basis: the mapping f may be chosen as we please. $\qquad \square$

Problem 3.19 Let G be an Abelian group generated by $\{t_1, t_2, \ldots, t_n\}$. Let F be a free Abelian group of rank n. Prove that F may be mapped epimorphically onto G.

Solution. By assumption F has a free basis $X = \{x_1, x_2, \ldots, x_n\}$ (say). Define $f : X \to G$ by $f(x_i) = t_i$ $(i = 1, 2, \ldots, n)$. Then f extends to a homomorphism $g : F \to G$ and g is surjective since every element of G has the form $t_1^{a_1} t_2^{a_2} \ldots t_n^{a_n} = g(x_1^{a_1} x_2^{a_2} \ldots x_n^{a_n})$. $\qquad \square$

Problem 3.20 Let F be the free Abelian group with free basis $\{x, y, z\}$. Let H be the subgroup of F generated by $\{x^2, x^{-1}y, yz^3\}$. Prove that F/H is cyclic of order 6.

Solution. Let G be the cyclic group of order 6 with generator t. We shall map F epimorphically onto G in such a way that H is the kernel. First we define $f : \{x, y, z\} \to G$ by $f(x) = t^3$, $f(y) = t^3$, $f(z) = t^{-1}$. Then f extends to an epimorphism, which we also denote by f, of F onto G. Let K be the kernel of f; we have to show that $H = K$. Let $u \in H$. Then, for suitable a, b, c, we have

$$u = (x^2)^a (x^{-1}y)^b (yz^3)^c = x^{2a-b} y^{b+c} z^{3c}$$

and thus, as t has order 6,
$$f(u) = [f(x)]^{2a-b}[f(y)]^{b+c}[f(z)]^{3c}$$
$$= (t^3)^{2a-b}(t^3)^{b+c}(t^{-1})^{3c}$$
$$= (t^6)^a = e.$$

Hence $H \subseteq K$. Conversely if $v \in K$ we have $v = x^m y^n z^p$ for suitable $m, n, p \in \mathbb{Z}$ and
$$e = f(v) = [f(x)]^m[f(y)]^n[f(z)]^p$$
$$= (t^3)^m(t^3)^n t^{-p}$$
$$= t^{3m+3n-p}.$$

This implies that 6 divides $3m+3n-p$ and so $p = 3q$ $(q \in \mathbb{Z})$ where $m+n-q$ is even, say $m+n-q = 2r$ $(r \in \mathbb{Z})$. Consequently we have
$$v = x^m y^n z^p = x^{2r-n+q} y^n z^{3q}$$
$$= (x^2)^r (x^{-1}y)^{n-q}(yz^3)^q$$
$$\in H.$$

Thus $K \subseteq H$. We now apply the first isomorphism theorem (Problem 2.70) to conclude that F/K is isomorphic to G. This establishes the result. $\square$

Problem 3.21 Let K be a subgroup of an Abelian group such that G/K is a finitely generated free Abelian group. Prove that there exists a subgroup F of G such that F is isomorphic to G/K and $G = K \times F$.

Solution. Let G/K have $\{x_1 K, x_2 K, \ldots, x_n K\}$ as a free basis. Then, as we say, we 'lift back' this free basis to G by letting F be the subgroup of G generated by $\{x_1, x_2, \ldots, x_n\}$. We prove first that $G = FK$. Let $x \in G$, then, for some $a_1, a_2, \ldots, a_n \in \mathbb{Z}$,
$$xK = (x_1 K)^{a_1}(x_2 K)^{a_2} \ldots (x_n K)^{a_n} = (x_1^{a_1} x_2^{a_2} \ldots x_n^{a_n})K$$
which implies that
$$x = x_1^{a_1} x_2^{a_2} \ldots x_n^{a_n} k$$
where $k \in K$. Thus $G = FK$. We now show that $F \cap K$ is trivial. Let $u \in F \cap K$, then, for some $b_1, b_2, \ldots, b_n \in \mathbb{Z}$, $u = x_1^{b_1} x_2^{b_2} \ldots x_n^{b_n}$ and so $K = uK = (x_1 K)^{b_1}(x_2 K)^{b_2} \ldots (x_n K)^{b_n}$. Since $\{u_1 K, u_2 K, \ldots, u_n K\}$ is a free basis we have $b_1 = b_2 = \ldots = b_n = 0$ and so $u = e$. Thus $F \cap K$ is trivial. We now have $G = F \times K$ and hence F is isomorphic to G/K.

The assertion does not necessarily hold if G/K is not a free group. $\square$

Problem 3.22 Let H be a subgroup of the Abelian group G such that H and G/H are finitely generated. Prove that G is finitely generated.

Solution. Let H be generated by $x_1, x_2, \ldots, x_m$ and let G/H be generated by $Hy_1, Hy_2, \ldots, Hy_n$. We claim that G is generated by $x_1, x_2, \ldots, x_m$, $y_1, y_2, \ldots, y_n$. Let $w \in G$. By assumption there exist $b_1, b_2, \ldots, b_n \in \mathbb{Z}$ such that

$$Hw = (Hy_1)^{b_1}(Hy_2)^{b_2} \ldots (Hy_n)^{b_n}$$

and so

$$w = y_1^{b_1} y_2^{b_2} \ldots y_n^{b_n} h$$

where $h \in H$. Since $h \in H$ there exist $a_1, a_2, \ldots, a_m \in \mathbb{Z}$ such that

$$h = x_1^{a_1} x_2^{a_2} \ldots x_m^{a_m}$$

and so

$$w = y_1^{b_1} y_2^{b_2} \ldots y_n^{b_n} x_1^{a_1} x_2^{a_2} \ldots x_m^{a_m}$$

which establishes the result.

The above conclusion is true without the assumption that G is commutative, only a slight modification of the above solution being necessary. $\square$

Problem 3.23 Let G be a finitely generated Abelian group and let H be a subgroup of G. Prove that H can be finitely generated.

Solution. If G is generated by one element then G is cyclic and so therefore is H. Let us argue by mathematical induction. We suppose that G is generated by $n\,(>1)$ elements $x_1, x_2, \ldots, x_n$ and that an Abelian group generated by less than n elements has the property that all its subgroups are finitely generated. Let K be the subgroup of G generated by x_1, $x_2, \ldots, x_{n-1}$. By the induction assumption, all subgroups of K are finitely generated. If $K = G$ then H is finitely generated as a subgroup of K. Suppose $K \neq G$. Then G/K is cyclic, being generated by Kx_n, and thus the subgroup HK/K of G/K is cyclic. By the second isomorphism theorem (Problem 2.74) $H/(H \cap K)$ is cyclic. Now $H \cap K$, being a subgroup of K, is finitely generated and hence (Problem 3.22) H is finitely generated. $\square$

Problem 3.24 Let T be the torsion subgroup of a finitely generated Abelian group G. Prove that T is finite.

Solution. By Problem 3.23, T is finitely generated by $t_1, t_2, \ldots, t_q$ (say) where, by assumption there exists r_i such that $t_i^{r_i} = e$ $(i = 1, 2, \ldots, q)$. This implies that every element of T is expressible as

$$t_1^{z_1} t_2^{z_2} \ldots t_q^{z_q}$$

where $0 \leqslant z_i \leqslant r_i$ $(i = 1, 2, \ldots, q)$. We deduce that T is finite. $\square$

Problem 3.25 Let p be a prime. Let G be an Abelian group and let G_p be the subset of G consisting of all elements the orders of which are powers of p. Prove that G_p is a subgroup of G. If G_p is finite show that $|G_p|$ is a power of p.

Solution. We observe first that $G_p \neq \varnothing$, for $e = e^{p^0} \in G_p$. Let $x, y \in G_p$. Then $x^{p^m} = y^{p^n} = e$ for some m, n. Hence

$$(xy)^{p^{m+n}} = x^{p^{m+n}}y^{p^{m+n}} = (x^{p^m})^{p^n}(y^{p^n})^{p^m} = ee$$
$$= e$$

and $(x^{-1})^{p^m} = (x^{p^m})^{-1} = e$. Thus G_p is a subgroup.

Let now G_p be a finite subgroup. Let $x \in G_p, x \neq e$ and let H be the subgroup generated by x. Suppose x has order p^r, then H has order p^r (see Problem 2.26). Consider G/H. The set of elements of G/H having orders which are powers of p is easily seen to be G_p/H. This observation affords a means for an induction argument. If $H = G_p$ then $|G_p| = p^r$ and if $H \neq G_p$ then $|G_p/H| < |G_p|$ and so we have the induction hypothesis that the assertion is true for G_p/H. Thus assume $|G_p/H| = p^s$ for some s and then

$$|G_p| = |G_p/H||H| = p^s p^r = p^{r+s}.$$

This completes the induction argument.

We call G_p, whether finite or not, the *p-Sylow subgroup* of G (after L. Sylow, Norwegian, 1832–1918). We notice that any subgroup of G consisting only of elements whose orders are powers of p is necessarily contained in G_p. If G_p is finite the order of such a subgroup is a power of p. $\square$

Problem 3.28 Prove that a finite Abelian group G has a nontrivial Sylow subgroup for each prime dividing $|G|$ and that G is the direct product of these Sylow subgroups.

Solution. We remark in passing that if p is a prime not dividing $|G|$ then the p-Sylow subgroup G_p is merely $\{e\}$.

Let $|G| = p_1^{a_1} p_2^{a_2} \ldots p_n^{a_n}$ be the factorisation of $|G|$ into the product of distinct primes $p_1, p_2, \ldots, p_n$. For ease of notation let G_i be the p_i-Sylow subgroup determined by p_i $(i = 1, 2, \ldots, n)$. We know that $|G_i|$ is a power of p_i and divides $|G|$. Thus $|G_i| = p_i^{b_i}$ where $b_i \leqslant a_i$ $(i = 1, 2, \ldots, n)$. We want to show that $|G_i| = p_i^{a_i}$ $(i = 1, 2, \ldots, n)$ and that $G = G_1 \times G_2 \times \ldots \times G_n$.

We employ arithmetical arguments. We let the integers q_i be defined by $p_i^{a_i} q_i = |G|$ $(i = 1, 2, \ldots, n)$. Since $p_i^{a_i}$ and q_i are relatively prime we have, by the division algorithm in $\mathbb{Z}$, that there exist $c_i, d_i \in \mathbb{Z}$ such that

$$c_i p_i^{a_i} + d_i q_i = 1 \quad (i = 1, 2, \ldots, n).$$

We also have that $q_1, q_2, \ldots, q_n$ have the greatest common divisor 1 and so, again by the division algorithm, there exist $f_1, f_2, \ldots, f_n \in \mathbb{Z}$ such that

$$f_1 q_1 + f_2 q_2 + \ldots + f_n q_n = 1.$$

Let us now prove that $G_i \cap G_1 G_2 \ldots G_{i-1} G_{i+1} \ldots G_n = \{e\}$. Let $x \in G_i \cap G_1 G_2 \ldots G_{i-1} G_{i+1} \ldots G_n$. Then $x = x_i = x_1 x_2 \ldots x_{i-1} x_{i+1} \ldots x_n$ where $x_j \in G_j$ $(j = 1, 2, \ldots, n)$. Since x_j to the power $p_j^{a_j}$ is e $(j = 1, 2, \ldots, n)$

we deduce that

$$(x_1 x_2 \ldots x_{i-1} x_{i+1} \ldots x_n)^{q_i} = e$$

and hence that

$$x = (x \text{ to the power } c_i p_i^{a_i} + d_i q_i)$$
$$= (x \text{ to the power } p_i^{a_i})^{c_i} (x^{q_i})^{d_i}$$
$$= e.$$

To prove that $G = G_1 G_2 \ldots G_n$ we let $y \in G$ and then

$$y = y^{f_1 q_1 + f_2 q_2 + \ldots + f_n q_n}$$
$$= (y^{q_1})^{f_1} (y^{q_2})^{f_2} \ldots (y^{q_n})^{f_n}.$$

But $(y^{q_i}$ to the power $p_i^{a_i}) = y^{|G|} = e$ and so $y^{q_i} \in G_i$ $(i = 1, 2, \ldots, n)$. Thus $y \in G_1 G_2 \ldots G_n$ and $G = G_1 G_2 \ldots G_n$. We have now established that $G = G_1 \times G_2 \times \ldots \times G_n$.

It is now immediately clear that $|G_i| = p_i^{a_i}$ since

$$p_1^{a_1} p_2^{a_2} \ldots p_n^{a_n} = |G| = |G_1||G_2| \ldots |G_n|$$
$$= p_1^{b_1} p_2^{b_2} \ldots p_n^{b_n}$$

yields $a_i = b_i$ $(i = 1, 2, \ldots, n)$. □

Problem 3.29 What are the orders of the Sylow subgroups of an Abelian group of order 2352?

Solution. Since $2352 = 2^4 \times 3 \times 7^2$ the Sylow subgroups have orders 16, 3, 49. □

Finitely generated Abelian groups arise in various algebraic and topological contexts; the structure and characterisation of such groups is known. If G is a finitely generated Abelian group then G is isomorphic to $T \times F$ where T is the torsion subgroup of G and F is a finitely generated free Abelian group isomorphic to G/T. (We already know that G determines T and that G/T is finely generated torsion-free; we are asserting a little more, namely that the concepts of free Abelian group and torsion-free Abelian group coincide for finitely generated Abelian groups—see Problem 3.21.) Furthermore G determines the rank of F and T is the direct product of its nontrivial Sylow subgroups, each of which is the direct product of cyclic groups of prime power order, the number and orders of which are unique. In other words a finitely generated Abelian group G is isomorphic to a direct product of the form

$$C_{p_1} s_1 \times C_{p_2} s_2 \times \ldots \times C_{p_m} s_m \times F_1 \times F_2 \times \ldots \times F_n \qquad (3.9)$$

where $C_{p_j} s_j$ is a cyclic group of order $p_j^{s_j}$ $(j = 1, 2, \ldots, m)$, $p_1, p_2, \ldots, p_m$ being not necessarily distinct primes, and where F_i is an infinite cyclic group ($i = 1, 2, \ldots, n$). If G is also isomorphic to

$$C_{p_1}\sigma_1 \times C_{p_2}\sigma_2 \times \ldots \times C_{p_{m'}}\sigma_{m'} \times F_1 \times F_2 \times \ldots \times F_{n'} \qquad (3.10)$$

then $m' = m$, $n' = n$ and with a possible renumbering of the subgroups $C_{p_j}\sigma_j$ is isomorphic to $C_{p_j}s_j$ ($j = 1, 2, \ldots, m$).

Problem 3.30 Find all Abelian groups of orders 4 and 6.

Solution. Strictly the phrase 'up to isomorphism' should be incorporated in the above as there is an infinity of groups of order 4. The phrase is commonly omitted, being understood to be implicit.

We have $4 = 2^2 = 2 \times 2$ and so (up to isomorphism) there are two Abelian groups of order 4, one being cyclic of order 2^2 and the other being the direct product of two cyclic groups of orders 2.

We have $6 = 2 \times 3$ and so there is (up to isomorphism) only one Abelian group of order 6, this being the direct product of a cyclic group of order 2 and a cyclic group of order 3. ☐

Problem 3.31 Find all Abelian groups of order 72.

Solution. Let C_r be a cyclic group of order r. Since $72 = 8 \times 9$, any Abelian group of this order is of one of the following types:

$$C_8 \times C_9, \; C_8 \times C_3 \times C_3, \; C_4 \times C_2 \times C_9, \; C_4 \times C_2 \times C_3 \times C_3,$$
$$C_2 \times C_2 \times C_2 \times C_9, \; C_2 \times C_2 \times C_2 \times C_3 \times C_3. \qquad ☐$$

Problem 3.32 Let G be a finitely generated free Abelian group and let H be a nontrivial subgroup of G. Prove that H is also a finitely generated free Abelian group.

Solution. Since H is a subgroup of G, H is torsion-free and finitely generated (Problem 3.23). By a remark above H is free Abelian. ☐

Problem 3.33 Give an example of an Abelian group that is not finitely generated.

Solution. The group of 3^n-roots of 1 given in Problem 2.25 is a torsion group but is not finite and so cannot be finitely generated (Problem 3.24). ☐

An Abelian group, particularly in applications, often appears in additive notation. In such an Abelian group A, the law of composition is *addition*. The identity element for this law of composition is the *zero of addition* 0 and the inverse of $x \in A$ is denoted by $-x$. For example $\mathbb{Z}$ with addition as composition is an additive group. We may rephrase our notions of Abelian groups in additive notation. The torsion subgroup T of A is $T = \{x : x \in A, mx = 0, m \in \mathbb{Z}, m \neq 0\}$. Any infinite cyclic group is isomorphic to $\mathbb{Z}$ and so a finitely generated free Abelian group is the *direct sum* of a finite number of isomorphic copies of $\mathbb{Z}$.

Problem 3.34 Let n be a positive integer and let K_n be the set of integers divisible by n. Prove that K_n is a subgroup of $\mathbb{Z}$ and that $\mathbb{Z}/K_n$ is a cyclic group of order n.

Solution. We have $K_n = \{x : x = ny, y \in \mathbb{Z}\}$ and K_n is a subgroup since

$$ny_1 + ny_2 = n(y_1 + y_2)$$
$$-(ny_1) = n(-y_1).$$

The cosets of K_n in $\mathbb{Z}$ are precisely K_n, K_n+1, K_n+2, $K_n+(n-1)$ where $K_n+r = \{u : u = k+r, k \in K\}$. It is not hard to see that K_n+1 (indeed any coset K_n+s where s and n are relatively prime) generates $\mathbb{Z}/K_n$.

We use the notation $\mathbb{Z}_n$ for $\mathbb{Z}/K_n$, $\mathbb{Z}_n$ is then the typical cyclic group of order n. □

Problem 3.35 Let B, C be subgroups of the additive Abelian group A. Prove that $B+C$ is a subgroup of A.

Solution. We have $B+C = \{x : x = b+c, b \in B, c \in C\}$, this is the smallest subgroup containing B and C. □

Instead of the (finite) direct product of groups we have now the direct sum, denoted by $\oplus$, of groups. Thus $\mathbb{Z}_{12} \oplus \mathbb{Z}_{12}$ is isomorphic to the direct sum of two cyclic groups each isomorphic to $\mathbb{Z}_{12}$. Every finitely generated additive Abelian group A is isomorphic to

$$\mathbb{Z}_{p_1}s_1 \oplus \mathbb{Z}_{p_2}s_2 \oplus \ldots \oplus \mathbb{Z}_{p_m}s_m \oplus \mathbb{Z} \oplus \mathbb{Z} \oplus \ldots \oplus \mathbb{Z} \qquad (3.11)$$

where n 'copies' of $\mathbb{Z}$ appear and where $p_1, p_2, \ldots, p_m$ are not necessarily distinct primes, this decomposition of A is unique up to isomorphism.

Problem 3.36 Find all Abelian groups of order 675.

Solution. Since $675 = 3^3 \times 5^2$ we have, up to isomorphism,

$$\mathbb{Z}_{27} \oplus \mathbb{Z}_{25}, \qquad \mathbb{Z}_{27} \oplus \mathbb{Z}_5 \oplus \mathbb{Z}_5$$
$$\mathbb{Z}_9 \oplus \mathbb{Z}_3 \oplus \mathbb{Z}_{25}, \qquad \mathbb{Z}_9 \oplus \mathbb{Z}_3 \oplus \mathbb{Z}_5 \oplus \mathbb{Z}_5$$
$$\mathbb{Z}_3 \oplus \mathbb{Z}_3 \oplus \mathbb{Z}_3 \oplus \mathbb{Z}_{25}, \qquad \mathbb{Z}_3 \oplus \mathbb{Z}_3 \oplus \mathbb{Z}_3 \oplus \mathbb{Z}_5 \oplus \mathbb{Z}_5.$$ □

EXERCISES

1. Let G be a cyclic group of order mn where m, n are integers whose greatest common divisor is 1. Prove that G has subgroups A, B where $|A| = m$, $|B| = n$ and that G is isomorphic to $A \times B$.

2. Let H_i be a subgroup of the group G_i $(i = 1, 2)$. Prove that $H_1 \times H_2$ is a subgroup of $G_1 \times G_2$. Suppose G_i is the cyclic group of order 4 generated by x_i $(i = 1, 2)$. Prove that $H = \{(x_1^m, x_2^m) : m = 0, 1, 2, 3\}$ is a subgroup of $G_1 \times G_2$ but is not of the form $H_1 \times H_2$ for any H_1, H_2.

3. Let G_i and H_i be groups and let $f_i : G_i \to H_i$ be a homomorphism with kernel K_i $(i = 1, 2)$. Prove that the mapping f defined by
$$f(g_1, g_2) = (f_1(g_1), f_2(g_2)) \quad (g_i \in G_i, i = 1, 2)$$
is a homomorphism with kernel $K_1 \times K_2$.

4. Let N_1, N_2 be normal subgroups of the group G. Prove that the mapping $x \to (xN_1, xN_2)$ is a homomorphism of G into $G/N_1 \times G/N_2$ with kernel $N_1 \cap N_2$.

5. Let G be the finite Abelian group generated by $x_1, x_2, \ldots, x_n$ where x_i has order n_i $(i = 1, 2, \ldots, n)$. Prove that $|G|$ divides $n_1 n_2 \ldots n_r$.

6. Let G be the free Abelian group of rank 2 having $\{a, b\}$ $(a, b \in G)$ as basis. Let $c = a^5 b^8$, $d = a^2 b^3$. Prove that $\{c, d\}$ is also a free basis of G.

7. Let G be the free (additive) Abelian group with free basis $\{a, b, c\}$. Let H be the subgroup generated by $3a$, $15b$, $7c$. Prove that H is free Abelian and that G/H is the direct sum of three cyclic groups of orders $3, 15, 7$.

8. Write down the array giving the law of (additive) composition in $\mathbb{Z}_6$. Find inverses of $2 + \mathbb{Z}_6$, $5 + \mathbb{Z}_6$.

9. Verify that $\mathbb{Z}_{36}$ is isomorphic to $\mathbb{Z}_4 \oplus \mathbb{Z}_9$.

10. Prove that $\mathbb{Z}_{20449}$ is the direct sum of its Sylow subgroups of orders 121 and 169.

Chapter 4

Symmetry, geometry

4.1 Permutation groups

Problem 4.1 Let X be a nonempty set and let $S(X)$ be the set of bijections of X onto X. Prove that $S(X)$ is a group under the circle composition of mappings.

Solution. By Problem 1.22 if f, g are bijections then $f \circ g$ is also a bijection. The circle-composition of mappings is associative (Problem 1.24). The identity mapping ι_X is trivially a bijection and (Problem 1.27) the inverse of a bijection is also a bijection. Hence $S(X)$ is a group.

A one–one mapping of a nonempty set X onto itself is often called a *permutation* of X. We have shown that the set $S(X)$ of all permutations of X is a group; this group is called the *symmetric group* on X. Any subgroup of $S(X)$ is called a *permutation group*. For notational brevity we usually omit the $\circ$ from $f \circ g$ in writing down the product of the permutations f and g. $\qquad\qquad\square$

An important case to be considered arises when the set X is finite.

Problem 4.2 Let X be a finite set of n elements. Prove that the order of the symmetric group on X is $n!$.

Solution. Let $X = \{x_1, x_2, \dots, x_n\}$. Let $p \in S(X)$. Then $p(x_1)$ may be any one of the n elements $x_1, x_2, \dots, x_n$; suppose $p(x_1) = x_{i_1}$. Since p is one–one and onto, $p(x_2)$ may then be any one of the $n-1$ elements $x_1, x_2, \dots, x_{i_1-1}, x_{i_1+1}, \dots, x_n$; suppose $p(x_2) = x_{i_2}$. Then $p(x_3)$ may be any one of the $n-2$ elements obtained by omitting x_{i_1}, x_{i_2} from $x_1, x_2, \dots, x_n$. Thus we have n choices for $p(x_1)$, $n-1$ choices for $p(x_2)$, $n-2$ choices for $p(x_3)$ and so on giving $n-(k-1)$ choices for $p(x_k)$. Thus, in all, there are $n(n-1)\dots2.1$ choices for p and this establishes the result. $\qquad\qquad\square$

Let X be finite and suppose $X = \{x_1, x_2, \dots, x_n\}$. Then, for a given permutation p of X it is sometimes convenient to write

$$p = \begin{pmatrix} x_1 & x_2 & \dots & x_n \\ p(x_1) & p(x_2) & \dots & p(x_n) \end{pmatrix} \qquad (4.1)$$

where the notation indicates that $p : x_i \to p(x_i)$ $(i = 1, 2, \dots, n)$. If we are considering abstractly some group of permutations it is usual to represent the permutations as being on the symbols $1, 2, \dots, n$ and so a permutation

r on $\{1, 2, \ldots, n\}$ is written as

$$r = \begin{pmatrix} 1 & 2 & \ldots & n \\ r(1) & r(2) & \ldots & r(n) \end{pmatrix}. \tag{4.2}$$

In this case we write S_n for the symmetric group of all permutations on the symbols 1, 2, ..., n; notice that in this instance the natural order of the symbols is irrelevant. The permutation $s \in S_4$ given by $s(1) = 2$, $s(2) = 4$, $s(3) = 3$, $s(4) = 1$ can be denoted in any of the following ways as

$$s = \begin{pmatrix} 1 & 2 & 3 & 4 \\ 2 & 4 & 3 & 1 \end{pmatrix} = \begin{pmatrix} 2 & 1 & 3 & 4 \\ 4 & 2 & 3 & 1 \end{pmatrix} = \begin{pmatrix} 4 & 3 & 1 & 2 \\ 1 & 3 & 2 & 4 \end{pmatrix} = \therefore \tag{4.3}$$

If $t \in S_4$ is given by $t(1) = 4$, $t(2) = 3$, $t(3) = 1$, $t(4) = 2$, i.e.

$$t = \begin{pmatrix} 1 & 2 & 3 & 4 \\ 4 & 3 & 1 & 2 \end{pmatrix}$$

the product ts is given by

$$1 \xrightarrow{s} 2 \xrightarrow{t} 3$$
$$2 \xrightarrow{s} 4 \xrightarrow{t} 2$$
$$3 \xrightarrow{s} 3 \xrightarrow{t} 1$$
$$4 \xrightarrow{s} 1 \xrightarrow{t} 4$$

$$\tag{4.4}$$

which we write succinctly as

$$ts = \begin{pmatrix} 1 & 2 & 3 & 4 \\ 4 & 3 & 1 & 2 \end{pmatrix}\begin{pmatrix} 1 & 2 & 3 & 4 \\ 2 & 4 & 3 & 1 \end{pmatrix} = \begin{pmatrix} 2 & 4 & 3 & 1 \\ 3 & 2 & 1 & 4 \end{pmatrix}\begin{pmatrix} 1 & 2 & 3 & 4 \\ 2 & 4 & 3 & 1 \end{pmatrix}$$

$$= \begin{pmatrix} 1 & 2 & 3 & 4 \\ 3 & 2 & 1 & 4 \end{pmatrix}. \tag{4.5}$$

The reader should note that another mutiplication convention is to be found in some textbooks.

Problem 4.3

Let $a = \begin{pmatrix} 1 & 2 & 3 & 4 & 5 & 6 & 7 \\ 4 & 6 & 7 & 3 & 5 & 1 & 2 \end{pmatrix}$

$b = \begin{pmatrix} 1 & 2 & 3 & 4 & 5 & 6 & 7 \\ 3 & 6 & 7 & 1 & 5 & 4 & 2 \end{pmatrix}.$

Find a^{-1} and verify that $a^{-1}ba \neq b$.

Solution. We may easily verify that

$$a^{-1} = \begin{pmatrix} 4 & 6 & 7 & 3 & 5 & 1 & 2 \\ 1 & 2 & 3 & 4 & 5 & 6 & 7 \end{pmatrix} = \begin{pmatrix} 1 & 2 & 3 & 4 & 5 & 6 & 7 \\ 6 & 7 & 4 & 1 & 5 & 2 & 3 \end{pmatrix}.$$

We have

$$ba = \begin{pmatrix} 1 & 2 & 3 & 4 & 5 & 6 & 7 \\ 3 & 6 & 7 & 1 & 5 & 4 & 2 \end{pmatrix}\begin{pmatrix} 1 & 2 & 3 & 4 & 5 & 6 & 7 \\ 4 & 6 & 7 & 3 & 5 & 1 & 2 \end{pmatrix}$$

$$ = \begin{pmatrix} 1 & 2 & 3 & 4 & 5 & 6 & 7 \\ 1 & 4 & 2 & 7 & 5 & 3 & 6 \end{pmatrix}$$

and hence

$$a^{-1}ba = \begin{pmatrix} 1 & 2 & 3 & 4 & 5 & 6 & 7 \\ 6 & 7 & 4 & 1 & 5 & 2 & 3 \end{pmatrix}\begin{pmatrix} 1 & 2 & 3 & 4 & 5 & 6 & 7 \\ 1 & 4 & 2 & 7 & 5 & 3 & 6 \end{pmatrix}$$

$$ = \begin{pmatrix} 1 & 2 & 3 & 4 & 5 & 6 & 7 \\ 6 & 1 & 7 & 3 & 5 & 4 & 2 \end{pmatrix} \neq b.$$

Notice that the inverse of the permutation a is the permutation that 'undoes' the effect of a. Thus, in general, if

$$p = \begin{pmatrix} 1 & 2 & \ldots & n \\ i_1 & i_2 & \ldots & i_n \end{pmatrix} \tag{4.6}$$

where $p(j) = i_j$ $(j = 1, 2, \ldots, n)$ we have

$$p^{-1} = \begin{pmatrix} i_1 & i_2 & \ldots & i_n \\ 1 & 2 & \ldots & n \end{pmatrix}. \qquad \square \tag{4.7}$$

Problem 4.4 Write down the elements of S_3 and find their orders.

Solution. S_3 has six elements, namely,

$$\begin{pmatrix} 1 & 2 & 3 \\ 1 & 2 & 3 \end{pmatrix}, \begin{pmatrix} 1 & 2 & 3 \\ 2 & 3 & 1 \end{pmatrix}, \begin{pmatrix} 1 & 2 & 3 \\ 3 & 1 & 2 \end{pmatrix},$$

$$\begin{pmatrix} 1 & 2 & 3 \\ 2 & 1 & 3 \end{pmatrix}, \begin{pmatrix} 1 & 2 & 3 \\ 3 & 2 & 1 \end{pmatrix}, \begin{pmatrix} 1 & 2 & 3 \\ 1 & 3 & 2 \end{pmatrix}, \begin{pmatrix} 1 & 2 & 3 \\ 1 & 2 & 3 \end{pmatrix}$$

is the identity permutation. $\begin{pmatrix} 1 & 2 & 3 \\ 2 & 3 & 1 \end{pmatrix}$ has order 3 for

$$\begin{pmatrix} 1 & 2 & 3 \\ 2 & 3 & 1 \end{pmatrix}\begin{pmatrix} 1 & 2 & 3 \\ 2 & 3 & 1 \end{pmatrix} = \begin{pmatrix} 1 & 2 & 3 \\ 3 & 1 & 2 \end{pmatrix} \neq \begin{pmatrix} 1 & 2 & 3 \\ 1 & 2 & 3 \end{pmatrix}$$

but

$$\begin{pmatrix} 1 & 2 & 3 \\ 2 & 3 & 1 \end{pmatrix}\begin{pmatrix} 1 & 2 & 3 \\ 2 & 3 & 1 \end{pmatrix}\begin{pmatrix} 1 & 2 & 3 \\ 2 & 3 & 1 \end{pmatrix} = \begin{pmatrix} 1 & 2 & 3 \\ 1 & 2 & 3 \end{pmatrix}.$$

Similarly $\begin{pmatrix} 1 & 2 & 3 \\ 3 & 1 & 2 \end{pmatrix}$ has order 3 and the three remaining permutations have order 2. $\qquad \square$

Problem 4.5 Prove that S_3 is isomorphic to the group of Problem 2.27.

Solution. It is a routine verification that the mapping u is an isomorphism where

$$u(e) = \begin{pmatrix} 1 & 2 & 3 \\ 1 & 2 & 3 \end{pmatrix}, \quad u(a) = \begin{pmatrix} 1 & 2 & 3 \\ 2 & 3 & 1 \end{pmatrix}, \quad u(b) = \begin{pmatrix} 1 & 2 & 3 \\ 3 & 1 & 2 \end{pmatrix}$$

$$u(c) = \begin{pmatrix} 1 & 2 & 3 \\ 2 & 1 & 3 \end{pmatrix}, \quad u(d) = \begin{pmatrix} 1 & 2 & 3 \\ 3 & 2 & 1 \end{pmatrix}, \quad u(f) = \begin{pmatrix} 1 & 2 & 3 \\ 1 & 3 & 2 \end{pmatrix}. \qquad \square$$

A nontrivial permutation p of $\{1, 2, \ldots, n\}$ is called a *cycle* of length r $(1 < r \leqslant n)$ if p has the form

$$p = \begin{pmatrix} i_1 & i_2 & \cdots & i_r & j_{r+1} & j_{r+2} & \cdots & j_n \\ i_2 & i_3 & \cdots & i_1 & j_{r+1} & j_{r+2} & \cdots & j_n \end{pmatrix}. \tag{4.8}$$

The identity permutation is called a cycle of length 1. The notation is commonly abbreviated to

$$p = (i_1 \, i_2 \ldots i_r) \tag{4.9}$$

where it is understood that $p(i_1) = i_2$, $p(i_2) = i_3, \ldots, p(i_r) = i_1$ and that the remaining symbols are unmoved by p. Thus we write

$$\begin{pmatrix} 1 & 2 & 3 & 4 & 5 & 6 & 7 & 8 \\ 1 & 8 & 5 & 4 & 6 & 2 & 7 & 3 \end{pmatrix} = (2 \quad 8 \quad 3 \quad 5 \quad 6). \tag{4.10}$$

Notice however that as this notation does not indicate the unmoved symbols, their existence and number must be inferred from the particular context, thus, for example, the cycle $(1 \quad 2 \quad 3)$ might be either $\begin{pmatrix} 1 & 2 & 3 \\ 2 & 3 & 1 \end{pmatrix}$ or $\begin{pmatrix} 1 & 2 & 3 & 4 & 5 \\ 2 & 3 & 1 & 4 & 5 \end{pmatrix}$. Conventionally the identity permutation is denoted by (1).

Problem 4.6 In S_7 verify that

$$(2 \quad 5 \quad 4)(3 \quad 6 \quad 7 \quad 5 \quad 1)(3 \quad 1 \quad 4) = (1 \quad 2 \quad 5)(4 \quad 6 \quad 7)$$
$$= (4 \quad 6 \quad 7)(1 \quad 2 \quad 5).$$

Solution. We may evaluate the products directly if we simply check carefully what happens to a given symbol. Thus the product

$$(2 \quad 5 \quad 4)(3 \quad 6 \quad 7 \quad 5 \quad 1)(3 \quad 1 \quad 4)$$

is obtained from

$$1 \to 4 \to 4 \to 2$$
$$2 \to 2 \to 2 \to 5$$
$$3 \to 1 \to 3 \to 3$$
$$4 \to 3 \to 6 \to 6$$
$$5 \to 5 \to 1 \to 1$$
$$6 \to 6 \to 7 \to 7$$
$$7 \to 7 \to 5 \to 4$$

and so

$$(2\ 5\ 4)(3\ 6\ 7\ 5\ 1)(3\ 1\ 4) = \begin{pmatrix} 1 & 2 & 3 & 4 & 5 & 6 & 7 \\ 2 & 5 & 3 & 6 & 1 & 7 & 4 \end{pmatrix}.$$

Rather more easily we establish that

$$(1\ 2\ 5)(4\ 6\ 7) = (4\ 6\ 7)(1\ 2\ 5) = \begin{pmatrix} 1 & 2 & 3 & 4 & 5 & 6 & 7 \\ 2 & 5 & 3 & 6 & 1 & 7 & 4 \end{pmatrix}.$$

We notice that there is no symbol moved by both $(1\ 2\ 5)$, $(4\ 6\ 7)$ and that these cycles commute. In general two cycles p, $r \in S_n$ are called *disjoint* if no symbol moved by p is moved by r and conversely. $\square$

Problem 4.7 Prove that disjoint cycles of S_n commute.

Solution. Let s, t be disjoint cycles of S_n. Let $s = (i_1\ i_2 \ldots i_u)$, $t = (j_1\ j_2 \ldots j_v)$ where, since s and t are disjoint,

$$\{i_1, i_2, \ldots, i_u\} \cap \{j_1, j_2, \ldots, j_v\} = \emptyset.$$

If $u + v < n$ let $w = n - u - v$ and let $k_1, k_2, \ldots, k_w$ be the remaining symbols.

For convenience let $i_{u+1} = i_1$ and then

$$(st)(i_a) = s(t(i_a)) = s(i_a) = i_{a+1}$$
$$(ts)(i_a) = t(s(i_a)) = t(i_{a+1}) = i_{a+1}.$$

Similarly

$$(st)(j_b) = (ts)(j_b)$$

and, trivially,

$$(st)(k_c) = (ts)(k_c).$$

Hence we have established that $st = ts$. $\square$

Problem 4.8 Express the permutation

$$\begin{pmatrix} 1 & 2 & 3 & 4 & 5 & 6 & 7 & 8 & 9 \\ 2 & 8 & 5 & 1 & 3 & 9 & 6 & 4 & 7 \end{pmatrix}$$

as a product of disjoint cycles.

Solution.

$$\begin{pmatrix} 1 & 2 & 3 & 4 & 5 & 6 & 7 & 8 & 9 \\ 2 & 8 & 5 & 1 & 3 & 9 & 6 & 4 & 7 \end{pmatrix} = (1\ 2\ 8\ 4)(3\ 5)(6\ 9\ 7).$$

It may be shown that every permutation is a product of disjoint cycles. ☐

A cycle of length 2 is called a *transposition*.

Problem 4.9 Express

$$p = \begin{pmatrix} 1 & 2 & 3 & 4 & 5 & 6 & 7 & 8 \\ 3 & 4 & 5 & 6 & 1 & 2 & 8 & 7 \end{pmatrix}$$ as a product of transpositions.

Solution. We write the permutation first as a product of disjoint cycles. Then

$$p = (1\ 3\ 5)(2\ 4\ 6)(7\ 8).$$

It is then immediate that

$$(1\ 3\ 5) = (1\ 5)(1\ 3), \qquad (2\ 4\ 6) = (2\ 6)(2\ 4)$$

and so

$$p = (1\ 5)(1\ 3)(2\ 6)(2\ 4)(7\ 8).$$

It may be shown that every permutation is a product of transpositions. ☐

Problem 4.10 In S_5 evaluate the product $(5\ 4)(1\ 2)(1\ 5)(2\ 4)$.

Solution. We have

$$1 \to 1 \to 5 \to 5 \to 4$$
$$2 \to 4 \to 4 \to 4 \to 5$$
$$3 \to 3 \to 3 \to 3 \to 3$$
$$4 \to 2 \to 2 \to 1 \to 1$$
$$5 \to 5 \to 1 \to 2 \to 2$$

i.e. $\qquad (5\ 4)(1\ 2)(1\ 5)(2\ 4) = \begin{pmatrix} 1 & 2 & 3 & 4 & 5 \\ 4 & 5 & 3 & 1 & 2 \end{pmatrix}.$ ☐

As well as permuting the symbols $x_1, x_2, \ldots, x_n$ or, equivalently, permuting the subscripts of the symbols we may also consider permutations of polynomials in these symbols viewed as commuting indeterminates.

Thus if $s = \begin{pmatrix} 1 & 2 & 3 \\ 3 & 1 & 2 \end{pmatrix}$ and $f(x_1, x_2, x_3) = x_1^4 x_2^5 + x_1^7 x_2^8 x_3^9$ the polynomial $(sf)(x_1, x_2, x_3)$ is to be

$$x_{s(1)}^4 x_{s(2)}^5 + x_{s(1)}^7 x_{s(2)}^8 x_{s(3)}^9 = x_3^4 x_1^5 + x_3^7 x_1^8 x_2^9$$

In general if $f(x_1,x_2,\ldots,x_n)$ is a polynomial in the n indeterminates $x_1,x_2,\ldots,x_n$ (over $\mathbb{Z}, \mathbb{Q}, \mathbb{R}$, etc., as the case may be) then $(sf)(x_1,x_2,\ldots,x_n)$ is to be the polynomial obtained by replacing x_1 by $x_{s(1)}$, x_2 by $x_{s(2)},\ldots,x_n$ by $x_{s(n)}$.

Problem 4.11 Let $s = \begin{pmatrix} 1 & 2 & 3 & 4 \\ 4 & 3 & 1 & 2 \end{pmatrix}$, $\qquad t = \begin{pmatrix} 1 & 2 & 3 & 4 \\ 2 & 1 & 4 & 3 \end{pmatrix}$.

Let $\qquad f(x_1,x_2,x_3,x_4) = x_1\,x_2^2\,x_3^3\,x_4^4 - x_1^3 + x_3^5 + 5x_1^7x_3^2x_4^8$.
Evaluate $(sf)(x_1,x_2,\ldots,x_n)$ and $(tf)(x_1,x_2,\ldots,x_n)$.

Solution.

$$(sf)(x_1,x_2,\ldots,x_n) = x_4x_3^2x_1^3x_2^4 - x_4^3 + x_1^5 + 5x_4^7x_1^2x_2^8$$
$$= x_1^3x_2^4x_3^2x_4 - x_4^3 + x_1^5 + 5x_1^2x_2^8x_4^7$$
$$(tf)(x_1,x_2,\ldots,x_n) = x_2x_1^2x_4^3x_3^4 - x_2^3 + x_4^5 + 5x_2^7x_4^2x_3^8$$
$$= x_1^2x_2x_3^4x_4^3 - x_2^3 + x_4^5 + 5x_2^7x_3^8x_4^2. \qquad \square$$

Problem 4.12 Let $f(x_1,x_2,x_3) = x_1^2+x_2^2+x_3^2$, let $g(x_1,x_2,x_3) = $ $= x_1\,x_2+x_2\,x_3+x_3\,x_1$ and let $h(x_1,x_2,x_3) = (x_1-x_2)(x_1-x_3)(x_2-x_3)$. Prove that, for all $s \in S_3$, $(sf)(x_1,x_2,x_3) = f(x_1,x_2,x_3)$, $sg(x_1,x_2,x_3) = g(x_1,x_2,x_3)$ and $(sh)(x_1,x_2,x_3) = \pm h(x_1,x_2,x_3)$.

Solution. We have $(sf)(x_1,x_2,x_3) = x_{s(1)}^2 + x_{s(2)}^2 + x_{s(3)}^2$. Since s is a permutation of 1, 2, 3 we have $x_{s(1)}^2 + x_{s(2)}^2 + x_{s(3)}^2 = x_1^2 + x_2^2 + x_3^2$ as required. Similarly $(sg)(x_1,x_2,x_3) = g(x_1,x_2,x_3)$. A permutation of x_1,x_2,x_3 permutes the factors x_1-x_2, x_1-x_3, x_2-x_3 with possible changes of sign. Thus

$$(sh)(x_1,x_2,x_3) = \pm h(x_1,x_2,x_3). \qquad \square$$

A polynomial $f(x_1,x_2,\ldots,x_n)$ such that $(sf)(x_1,x_2,\ldots,x_n) = f(x_1,x_2,\ldots,x_n)$ for all transpositions $s \in S_n$ is called *symmetric*. Since every permutation is a product of transpositions it is equivalent that $f(x_1,x_2,\ldots,x_n)$ is symmetric if and only if $(sf)(x_1,x_2,\ldots,x_n) = f(x_1,x_2,\ldots,x_n)$ for all transpositions $s \in S_n$. A polynomial $h(x_1,x_2,\ldots,x_n)$ such that $(sh)(x_1,x_2,\ldots,x_n) = -h(x_1,x_2,\ldots,x_n)$ for all transpositions $s \in S_n$ is called *skew-symmetric*. It is equivalent that $f(x_1,x_2,\ldots,x_n)$ is skew-symmetric if and only if $(sf)(x_1,x_2,\ldots,x_n) = (-1)^\varepsilon f(x_1,x_2,\ldots,x_n)$ for all $s \in S_n$ where $\varepsilon = +1$ if s is a product of an odd number of transpositions and $\varepsilon = 2$ if s is the product of an even number of transpositions.

Problem 4.13 Let $h(x_1,x_2,x_3) = (x_1-x_2)(x_1-x_3)(x_2-x_3)$. Prove that $A_3 = \{s : s \in S_3, \ (sh)(x_1, x_2, x_3) = h(x_1, x_2, x_3)\}$ is a normal subgroup of S_3 of index 2.

Solution. It may be verified directly that

$$A_3 = \left\{ \begin{pmatrix} 1 & 2 & 3 \\ 1 & 2 & 3 \end{pmatrix}, \begin{pmatrix} 1 & 2 & 3 \\ 2 & 3 & 1 \end{pmatrix}, \begin{pmatrix} 1 & 2 & 3 \\ 3 & 1 & 2 \end{pmatrix} \right\} \qquad \square$$

which proves the result.

Problem 4.14 Let $h(x_1, x_2, \ldots, x_n)$ be the polynomial

$$\prod_{\substack{i,j=1 \\ i<j}}^{n} (x_i - x_j) = (x_1 - x_2)(x_1 - x_3)\ldots(x_1 - x_n)$$
$$\times (x_2 - x_3)\ldots(x_2 - x_n) \qquad (4.11)$$
$$\times (x_{n-1} - x_n).$$

If t is a transposition of S_n prove that

$$(th)(x_1, x_2, \ldots, x_n) = -h(x_1, x_2, \ldots, x_n)$$

and deduce that $A_n = \{s : s \in S_n, \ (sh)(x_1, x_2, \ldots, x_n) = h(x_1, x_2, \ldots, x_n)\}$ is a normal subgroup of S_n of index 2.

Solution. Let t be the transposition $(a\ b)$ where we may suppose $a < b$.

Let us determine the effect of t on $h(x_1, x_2, \ldots, x_n)$ by considering the effect of t on a typical factor $x_i - x_j$. If neither i nor j is a or b then $x_i - x_j$ is unaltered by t. Suppose either i or j is a or b but a and b do not both occur. Then $h(x_1, \ldots, x_n)$ has the pair of factors $x_i - x_a$ and $x_i - x_b$ $(i < a)$, $x_a - x_i$ and $x_i - x_b$ $(a < i < b)$ and $x_a - x_i$ and $x_b - x_i$ $(b < i)$; in each case the product of the two factors of the pair is unaltered by t. The factor remaining to be considered is $x_a - x_b$ and this is altered in sign by t. Thus finally $(t)(x_1, x_2, \ldots, x_n) = -h(x_1, x_2, \ldots, x_n)$.

It is easy to show, by Problem 2.31, that A_n is a subgroup. We now prove that $|S_n : A_n| = 2$. Let $c = (1\ 2)$, then $(ch)(x_1, x_2, \ldots, x_n) = -h(x_1, x_2, \ldots, x_n)$. Let $s \in S_n$. Either $(sh)(x_1, x_2, \ldots, x_n) = -h(x_1, x_2, \ldots, x_n)$ or $(sh)(x_1, x_2, \ldots, x_n) = -h(x_1, x_2, \ldots, x_n)$. In the first instance $s \in A_n$ and in the second instance

$$\begin{aligned}
((c^{-1}s)h)(x_1, x_2, \ldots, x_n) &= ((cs)h(x_1, x_2, \ldots, x_n)) \\
&= (c(sh))(x_1, x_2, \ldots, x_n) \\
&= (c(-h(x_1, x_2, \ldots, x_n))) \\
&= -(ch)(x_1, x_2, \ldots, x_n) \\
&= h(x_1, x_2, \ldots, x_n)
\end{aligned}$$

and thus $c^{-1}s \in A_n$ or $s \in cA_n$. Hence $S_n = A_n \cup cA_n$ and this establishes the result.

74

A permutation s such that $(sh)(x_1, x_2, ..., x_n) = h(x_1, x_2, ..., x_n)$ is called an *even permutation*, otherwise s is called an *odd permutation*. A_n is the set of all even permutations and is called the *alternating group on n symbols*. A_n has order $\frac{1}{2}n!$ and consists precisely of those permutations expressible as products of even numbers of transpositions.   □

Problem 4.15 Write down the elements of A_4 and prove that there are four of order 2.

Solution. In cycle notation we have, (1), (1 2 3), (1 3 2), (1 2 4), (1 4 2), (2 3 4), (2 4 3), (1 3 4), (1 4 3), (1 2) (3 4), (1 3) (2 4), (1 4) (2 3), of which the last three have order 2. Indeed these elements, together with (1) form a normal subgroup of A_4 isomorphic to the Klein four-group.   □

Problem 4.16 Let G be a group and let H be a subgroup of index n. Let $G = g_1 H \cup g_2 H \cup ... \cup g_n H$ be a coset decomposition. For each $g \in G$ let $p(g)$ be the permutation of the cosets $g_1 H, g_2 H, ..., g_n H$ given by

$$p(g) = \begin{pmatrix} g_1 H & g_2 H & \cdots & g_n H \\ gg_1 H & gg_2 H & \cdots & gg_n H \end{pmatrix}.$$

Prove that p is a homomorphism from G into the symmetric group S_n on the symbols $g_1 H, g_2 H, ..., g_n H$ and that p has kernel K where

$$K = g_1 H g_1^{-1} \cap g_2 H g_2^{-1} \cap ... \cap g_n H g_n^{-1}.$$

Solution. We first observe that $p(g)$ is genuinely a permutation since, first, $gg_i H$ is a coset and, second, $gg_i H = gg_j H$ if and only if $i = j$. To prove that p is a homomorphism let $x, y \in G$, then

$$p(xy) = \begin{pmatrix} g_1 H & g_2 H & \cdots & g_n H \\ xyg_1 H & xyg_2 H & \cdots & xyg_n H \end{pmatrix}$$

$$= \begin{pmatrix} yg_1 H & yg_2 H & \cdots & yg_n H \\ xyg_1 H & xyg_2 H & \cdots & xyg_n H \end{pmatrix} \begin{pmatrix} g_1 H & g_2 H & \cdots & g_n H \\ yg_1 H & yg_2 H & \cdots & yg_n H \end{pmatrix}$$

$$= p(x)p(y).$$

Also
$$\begin{aligned} K &= \{x \in G : p(x) = \iota\} \\ &= \{x \in G : xg_i H = g_i H, \quad i = 1, 2, ..., n\} \\ &= \{x \in G : g_i^{-1} x g_i H = H, \quad i = 1, 2, ..., n\} \\ &= \{x \in G : g_i^{-1} x g_i \in H, \quad i = 1, 2, ..., n\} \\ &= \{x \in G : x \in g_i H g_i^{-1}, \quad i = 1, 2, ..., n\} \\ &= \bigcap_{i=1}^{n} g_i H g_i^{-1}. \end{aligned}$$

If G is a finite group and if H is trivial then K is also trivial and G is then

75

F

mapped monomorphically into a permutation group $S_n (n = |G|)$ on the elements of G as symbols. This result is known as Cayley's theorem (after A. Cayley, English, 1801–1895). □

In the above we see that $p(G)$, as a subgroup of S_n, is a permutation group on $\{g_1 H, g_2 H, \ldots, g_n H\}$. It is convenient to say that G 'acts' as a permutation group on $\{g_1 H, g_2 H, \ldots, g_n H\}$. More generally if X is a nonempty set then we say that a group G acts as a permutation group on X if there exists a homomorphism $p : G \to S(X)$. For $g \in G$, $x \in X$ we define

$$g(x) = p(g)(x),$$

$p(g)$ being the permutation induced by $g \in G$.

Problem 4.17 Let G be a group and let H be a subgroup of index n. Prove that H contains a normal subgroup K of G such that $|G:K| \leqslant n!$.

Solution. By the previous problem G is mapped homomorphically by p onto $p(G)$ which is a subgroup of S_n, the symmetric group on the cosets of H in G. Letting, as above, K be the kernal of p, then $K \subseteq H$ and G/K is isomorphic to $p(G)$. But $|p(G)| \leqslant |S_n| = n!$ and this proves the result. □

4.2 Orbits, stabilisers

Let G be a group acting as a group of permutations of a nonempty set X, then for each $g \in G$ the mapping $x \to g(x)$ $(x \in X)$ is a bijection.

Problem 4.18 Let G be a permutation group on a nonempty set X. Define a relation $\sim$ on X by $x \sim y$ $(x, y \in X)$ if and only if $x = f(y)$ for some $f \in G$. Prove that $\sim$ is an equivalence relation on X.

Solution. We have $x \sim x$ $(x \in X)$ since $x = \iota_X(x)$ where ι_X is the identity permutation on X. If $x \sim y$ $(x, y \in X)$ then $x = g(y)$ for some $g \in G$ and so, as $g^{-1}(x) = g^{-1}(g(y)) = g^{-1}g(y) = \iota_X(y) = y$, we have $y \sim x$. If $x \sim y$ and $y \sim z$ $(x, y, z \in X)$ then $x = g(y)$, $y = h(z)$ $(g, h \in G)$ and we have $x \sim z$ since $x = g(y) = g(h(z)) = (gh)(z)$. Thus $\sim$ is an equivalence relation.

The equivalence classes are called *orbits*, the orbit $G(x)$ determined by $x \in X$ being

$$G(x) = \{y : y = g(x), \quad g \in G\}. \qquad \square \quad (4.12)$$

A group G of permutations on a set X is said to act *transitively* on X or to be a *transitive* permutation group on X if for some $x \in X$, $G(x) = X$. If G acts transitively on X then, in fact, for all $y \in X$, $G(y) = X$. Notice that this usage of the word 'transitive' is distinct from the earlier usage in discussing equivalence relations (See Problem 1.30 and succeeding remarks).

Problem 4.19 Let G be a permutation group on a set X and let $x \in X$. Prove that $H = \{g : g \in G, g(x) = x\}$ is a subgroup of G.

Solution. Clearly $\iota_X \in H$ and if $g, h \in H$ we have $(gh)(x) = g(h(x)) = g(x) = x$ and $g^{-1}(x) = g^{-1}(g(x)) = (g^{-1}g)(x) = \iota_X x = x$, hence H is a subgroup of G.

We call $\{g : g \in G, \ g(x) = x\}$ the *stabiliser* of x, written $\text{Stab}_G(x)$. ☐

Problem 4.20 Let G be a permutation group on a finite set X and let $x \in X$. Prove that the number of elements in $G(x)$ is precisely $|G : \text{Stab}_G(x)|$.

Solution. The reader should compare this Problem with Problem 2.53. Let $|G : \text{Stab}_G(x)| = n$. Then, for suitable $c_1, c_2, \ldots, c_n \in G$ we have

$$G = c_1 \text{Stab}_G(x) \cup c_2 \text{Stab}_G(x) \cup \ldots \cup c_n \text{Stab}_G(x).$$

We wish to show that $c_1(x), c_2(x), \ldots, c_n(x)$ are n distinct elements of X and that $G(x) = \{c_1(x), c_2(x), \ldots, c_n(x)\}$. First, if for some i, j we have $c_i(x) = c_j(x)$ then $(c_j^{-1}c_i)(x) = c_j^{-1}(c_i(x)) = c_j^{-1}(c_j(x)) = (c_j^{-1}c_j)(x) = x$ and so $c_j^{-1}c_i \in \text{Stab}_G(x)$. This implies, by Problem 2.46, that $c_j \text{Stab}_G(x) = c_i \text{Stab}_G(x)$ and so $i = j$. Second, if $g \in G$ then for some k ($1 \leqslant k \leqslant n$) and some $b \in \text{Stab}_G(x)$ we have $g = c_k b$ and this implies that $g(x) = (c_k b)(x) = c_k(b(x)) = c_k(x)$.

Notice that if $X = G$ and if we define the action of G in G by $g(x) = g^{-1}xg$ ($x \in G$) then G is a permutation group on G, the orbits being the conjugacy classes. In this case $\text{Stab}_G(x)$ is the centraliser $C_G(x)$ of x in G and we obtain the known result that the number of elements in the conjugacy class containing x is $|G : C_G(x)|$. ☐

4.3 Symmetry groups Some groups arise geometrically as groups of symmetry of regular plane and solid figures. If A, B, C are the vertices of an equilateral triangle with centroid O, an anticlockwise rotation about O through an angle of $\frac{2}{3}\pi$ in the plane of the triangle moves the vertices in accordance with the permutation $s = \begin{pmatrix} A & B & C \\ B & C & A \end{pmatrix}$ which may be represented diagramatically as in Figure 4.1.

Similarly a reflection about the line AO moves the vertices in accordance with the permutation $t = \begin{pmatrix} A & B & C \\ A & A & B \end{pmatrix}$ as in Figure 4.2.

In both cases only the labelling of the vertices shows that the triangle has been moved, each movement corresponding to a symmetry of the triangle and conversely. The set of all such movements of the triangle, the law of composition being the act of performing one movement after another, is a group, the so-called *symmetry group of the equilateral triangle*.

77

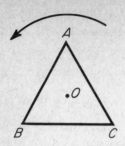
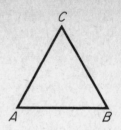

Figure 4.1

Problem 4.21 In the above prove that $ts = s^2t$ and interpret geometrically. What is the symmetry group of the equilateral triangle?

Solution. We may verify that, as permutations,

$$ts = \begin{pmatrix} A & B & C \\ C & B & A \end{pmatrix} = s^2t.$$

We have that ts is an anticlockwise rotation about O through an angle $\frac{2}{3}\pi$ followed by a reflection about an axis fixed in the initial position of AO, A itself having moved to the previous position of B. $\begin{pmatrix} A & B & C \\ C & B & A \end{pmatrix}$ is a reflection about an axis fixed in the initial position of BO. s^2t is a reflection about an axis in the initial position of OA followed by an anticlockwise notation through an angle of $\frac{4}{3}\pi$. Geometrically, ts, $\begin{pmatrix} A & B & C \\ C & B & A \end{pmatrix}$ and s^2t are one and the same movement of the triangle.

Each symmetry of the triangle is either a rotation about O through a multiple of $\frac{2}{3}\pi$ or is a reflection about one of the three axes AO, BO or CO, and each symmetry is a permutation of A, B, C. Conversely each permutation of A, B, C yields a symmetry of the triangle and so the symmetry group

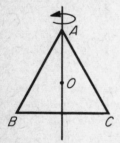
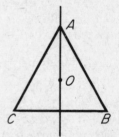

Figure 4.2

78

of the triangle is the group of all permutations on A, B, C and so is isomorphic to S_3. □

We may make some of the above notions more precise. The Euclidean plane $\mathbb{R}^2$ is the Cartesian product $\mathbb{R} \times \mathbb{R}$ with the usual idea of distance, that is, if P, Q are points of $\mathbb{R}^2$ with coordinates (x_P, y_P), (x_Q, y_Q) respectively and if $d(P, Q)$ is their distance apart we have

$$d(P, Q) = \sqrt{[(x_P - x_Q)^2 + (y_P - y_Q)^2]}. \tag{4.13}$$

A mapping $f : \mathbb{R}^2 \to \mathbb{R}^2$, such that f preserves the distance between any two points, is called an *isometry* (Greek *metron* $=$ measure); thus $f : \mathbb{R}^2 \to \mathbb{R}^2$ is an isometry if and only if $d(f(P), f(Q)) = d(P, Q)$ for all P, $Q \in \mathbb{R}^2$. It can be shown that an isometry f is both one–one and onto and that f is determined completely by its action on the vertices of any given triangle. On the other hand if f is a one–one distance-preserving mapping of any set of three noncollinear points into $\mathbb{R}^2$ then it can be shown that f extends uniquely to an isometry of $\mathbb{R}^2$. As examples of isometries of $\mathbb{R}^2$ we have the mappings $g_{a,b}$, r_α, t given by

$$g_{a,b}(x, y) = (x + a, y + b)$$
$$r_\alpha(x, y) = (x \cos \alpha + y \sin \alpha, -x \sin \alpha + y \cos \alpha) \tag{4.14}$$
$$t(x, y) = (x, -y) \quad (x, y \in \mathbb{R})$$

$g_{a,b}$ being a uniform displacement, r_α being a rotation about the origin O through an angle and t being a reflection about the x-axis. An isometry f is said to be a *symmetry transformation* or a *symmetry* of a plane figure F if $f(F) = F$.

Problem 4.22 Prove that the set T of isometries of $\mathbb{R}^2$ is a permutation group on $\mathbb{R} \times \mathbb{R}$ and that the set of symmetries of a plane figure F forms a subgroup $S(F)$ of T.

Solution. Let f, g be isometries. Then fg is an isometry since, for $P, Q \in \mathbb{R}^2$, we have

$$d((fg)(P), (fg)(Q)) = d(f(g(P)), f(g(Q)))$$
$$= d(g(P), g(Q))$$
$$= d(P, Q)$$

and, as f is a bijection, a mapping f^{-1} exists (Problem 1.27) and f^{-1} is an isometry since

$$d(f^{-1}(P), f^{-1}(Q)) = d(f(f^{-1}(P)), f(f^{-1}(Q)))$$
$$= d((ff^{-1})(P), (ff^{-1})(Q))$$
$$= d(P, Q).$$

Thus T is a group and it is now immediate that $S(F)$ is a subgroup since $S(F) = \text{Stab}_T(F)$.

We call $S(F)$ the *symmetry group of* F. To find $S(F)$ when F has at least three noncollinear points, it is sufficient, by the above remarks, to consider all distance-preserving (shape-preserving) mappings of F into F since each such mapping extends uniquely to an isometry of $\mathbb{R}^2$. ☐

Problem 4.23 Let F be an isosceles but not equilateral triangle. Find $S(F)$.

Solution. Let F be the triangle with vertices A, B, C in which $AB = AC$ and $AB \neq BC$ (see Figure 4.3).

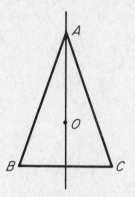

Figure 4.3

The only nontrivial symmetry is a reflection about the axis AO where O is the centroid. Thus $S(F)$ is the cyclic group of order 2. ☐

Problem 4.24 Let C be a circle. Find $S(C)$.

Solution. For convenience suppose the centre of C is the origin O and that the radius of C is 1.

If we regard C as being a solid wheel in a horizontal plane with a vertical axle through O we realise that any symmetry of C is either a turning about the axle or is a turning about the axle followed by a turning of the wheel upside down. More precisely any symmetry is either a rotation about O or is a rotation about O followed by a reflection in a diameter. Let r_α be a rotation about O through an angle α and let t_β be a reflection about a diameter making an angle β with the coordinate axis Ox. Denoting a typical point on the circumference of C by P_θ where θ is the polar angle we have

$$P_\theta = P_{\theta + 2n\pi} \quad (n = 0, \pm 1, \dots) \tag{4.15}$$

and we deduce from the diagrams in Figure 4.4 that

$$r_\alpha(P_\theta) = P_{\theta+\alpha}$$
$$t_\beta(P_\theta) = P_{2\beta-\theta}.$$

(4.16)

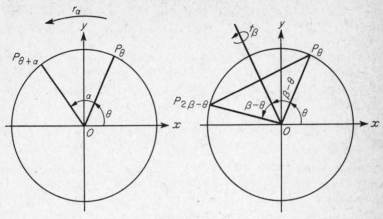

Figure 4.4

The subgroup $R(C)$ of $S(C)$ consisting of the rotations is an Abelian subgroup since

$$r_\alpha r_\beta = r_{\alpha+\beta} = r_{\beta+\alpha} = r_\beta r_\alpha.$$

Indeed if $\mathbb{R}^+$ denotes the additive group of real numbers and if $2\pi\mathbb{Z}$ is the subgroup of $\mathbb{R}^+$ consisting of all integral multiples of 2π then the mapping $h : R(C) \to \mathbb{R}^+/2\pi\mathbb{Z}$ given by $h(r_\alpha) = \alpha + 2\pi\mathbb{Z}$ is an isomorphism; in particular, we observe that r_α has finite order if and only if α is of the form $2\pi q$ where $q \in \mathbb{Q}$.

We now claim that $R(C)$ is a normal subgroup of $S(C)$ of index 2 and we establish this by showing that $S(C) = R(C) \cup t_0 R(C)$. Let $f \in S(C)$ and suppose $f \in R(C)$. By the remarks above $f = t_\beta r_\alpha$ for some angles α, β. But, since

$$(t_\beta r_\alpha)(P_\theta) = t_\beta(r_\alpha(P_\theta)) = t_\beta(P_{\theta+\alpha}) = P_{2\beta-\theta-\alpha} = t_0(P_{\theta+\alpha-2\beta})$$
$$= (t_0 r_{\alpha-2\beta})(P_\theta)$$

we have that $f = t_\beta r_\alpha = t_0 r_{\alpha-2\beta} \in t_0 R(c)$ and so we deduce that $S(C) = R(C) \cup t_0 R(C)$; geometrically this last equation implies that every reflection t_β is the product of a suitable rotation and the reflection t_0. Since $R(C)$ is a subgroup of $S(C)$ of index 2 it follows that $R(C)$ is a normal subgroup of $S(C)$ (Problem 2.51).

We further observe that if $2\gamma = 2\beta - \alpha$ then $t_\beta r_\alpha' = t_\gamma$ and so we conclude that every symmetry of C is either a rotation or a reflection. □

81

Problem 4.25 In the notation of the previous problem prove that $t_\alpha^{-1} r_\beta t_\alpha = r_\beta^{-1}$ and interpret geometrically.

Solution. We observe that $t_\alpha^{-1} = t_\alpha$ and that $r_\beta^{-1} = r_{-\beta}$. Thus
$$(t_\alpha^{-1} r_\beta t_\alpha)(P_\theta) = (t_\alpha^{-1} r_\beta)(t_\alpha(P_\theta)) = (t_\alpha^{-1} r_\beta)(P_{2\alpha-\theta}) = t_\alpha(r_\beta(P_{2\alpha-\theta}))$$
$$= t_\alpha(P_{2\alpha+\beta-\theta}) = P_{2\alpha-(2\alpha+\beta-\theta)} = P_{\theta-\beta} = r_{-\beta}(P_\theta) = r_\beta^{-1}(P_\theta)$$
and so we have $t_\alpha^{-1} r_\beta t_\alpha = r_\beta^{-1}$.

Geometrically a reflection about a line at an angle α to the x-axis, followed by a rotation through an angle β and followed by a reflection about a line at an angle α to the x-axis is equivalent to a rotation through an angle $-\beta$. $\qquad\square$

Problem 4.26 Find the symmetry group $S(L_n)$ of a regular polygon L_n of n sides.

Solution. For convenience we may suppose that L_n is inscribed in a circle C with centre $(0,0)$ and radius 1 and that the vertices of L_n occur at the points P_θ where $\theta = 2k\pi/n$ ($k = 0, 1, \ldots, n-1$). We illustrate the general situation by means of a regular pentagon (Figure 4.5).

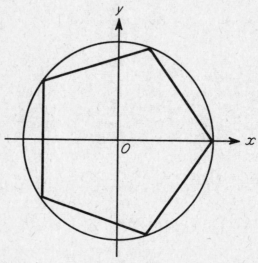

Figure 4.5

Every symmetry of L_n extends to an isometry of $\mathbb{R}^2$ which yields, in its turn, a symmetry of C but, on the other hand, not conversely since we may only rotate through angles which are multiples of $2\pi/n$. Thus if $R(L_n)$ denotes the subgroup of $S(L_n)$ consisting of rotations then $R(L_n)$ is a

82

cyclic group of order n generated by $r_{2\pi/n}$. The reflection t_0 is a symmetry of L_n and we deduce, as in the previous problem, that $S(L_n) = R(L_n) \cup t_0 R(L_n)$.

The group $S(L_n)$ is said to be *generated* by $r_{2\pi/n}$, t_0 with the *relations* $(r_{2\pi/n})^n = \iota_{L_n}$, $t_0^2 = \iota_{L_n}$, $t_0^{-1} r_{2\pi/n} t_0 = r_{2\pi/n}^{-1}$, ι_{L_n} being the identity symmetry on L_n. The group $S(L_n)$ is called the *dihedral group* of order $2n$.

Abstractly a group G is generated by a subset $\{g_1, g_2, \ldots, g_m\}$ of G if G is the least subgroup of G containing the subset. Somewhat imprecisely a set of equations in the generators $g_1, g_2, \ldots, g_m$ is said to be a *set of relations* if every equation in G involving $g_1, g_2, \ldots, g_m$ is a consequence of the set of relations. Thus the abstract dihedral group D_n is the group generated by a, b with the relations $a^n = e$, $b^2 = e$, $b^{-1} ab = a^{-1}$. $\square$

If, now, we go on to consider isometries in Euclidean space $\mathbb{R}^3$ we have appropriate extensions of the above ideas and results; it can be shown that an isometry of $\mathbb{R}^3$ is uniquely determined by the effect of the isometry on the four vertices of any tetrahedron. In $\mathbb{R}^3$ the uniform displacement $g_{a,b,c}$ is given by

$$g_{a,b,c}(x, y, z) = (x+a, y+b, z+c) \quad (x, y, z \in \mathbb{R}).$$

A rotation r_ω is an isometry for which there is a line the points of which remain fixed under the isometry; for a nonidentity rotation this line is unique and is called the *axis of the rotation*, all points not on the fixed line are rotated through a given angle called the *angle of the rotation*. If the axis of a rotation passes through the origin O then this axis intersects the surface of the sphere, centred at O and with radius 1, in two points called *poles*; the poles are the only points of the surface of the sphere unmoved by the rotation. Notice, however, that two distinct rotations may have the same poles, only the angles of the rotations being different.

Problem 4.27 Let G be a group of rotations the axes of which pass through the origin O. Let P be a pole of a nonidentity rotation r of G and let $g \in G$. Prove that $Q = g(P)$ is a pole of the rotation grg^{-1}.

Solution. We note first that as distances are preserved under rotations Q is a point of the sphere centred at O with radius 1. Now grg^{-1} is not the identity rotation and $(grg^{-1})(Q) = (gr)(g^{-1}(Q)) = (gr)(P) = g(r(P)) = g(Q) = Q$. Thus, as Q is fixed by grg^{-1}, Q is a pole of grg^{-1}.

An immediate consequence of this result is that a group G of rotations acts as a permutation group on the set of the poles of the rotations in G. $\square$

As well as the isometries of uniform displacements and rotations in $\mathbb{R}^3$ we also have reflections in planes, the reflection t given by

$$t(x, y, z) = (x, y, -z)$$

is the reflection in the plane $z = 0$. We shall confine ourselves in the following discussion mainly to the group of rotational symmetries of a regular polyhedron N; this group is denoted by $S_R(N)$ and is, of course, a subgroup of the full symmetry group, $S(N)$, of N.

Problem 4.28 Let T be a regular tetrahedron (Figure 4.6). Find $S(T)$.

Solution. We know that T has four faces, each being an equilateral triangle, and we may suppose that T is inscribed in the sphere with centre at O and radius 1; O is then the centroid of T. Let T have vertices A, B, C, D. $S(T)$ is then a permutation group on A, B, C, D.

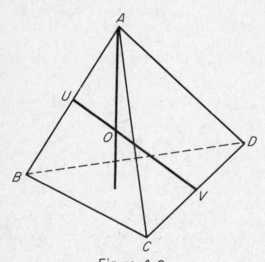

Figure 4.6

We consider two obvious symmetries of T. We may rotate T about the axis AO through angles $\frac{2}{3}\pi, \frac{4}{3}\pi$ to obtain symmetries of T; thus the permutations $\begin{pmatrix} A & B & C & D \\ A & C & D & B \end{pmatrix}, \begin{pmatrix} A & B & C & D \\ A & D & B & C \end{pmatrix}$ belong to $S(T)$. In cycle notation $(B\,C\,D), (B\,D\,C) \in S(T)$. If U, V are the midpoints of AB, CD respectively then UV passes through O and is an axis of rotational symmetry of T. In this case we have the permutation $\begin{pmatrix} A & B & C & D \\ B & A & D & C \end{pmatrix} = (A\,B)(C\,D)$.

By similar arguments we see that $S(T)$ contains the permutations $(A\,C\,D)$, $(A\,D\,C)$, $(A\,B\,D)$, $(A\,D\,B)$, $(A\,B\,C)$, $(A\,C\,B)$, $(A\,C)(B\,D)$, $(A\,D)(B\,C)$. Furthermore, there are no other nontrivial symmetries of T, rotational or otherwise, and so we conclude that $S(T)$ is the alternating group of order 12 on A, B, C, D.

The group $S(T)$ is sometimes called the *tetrahedral group*. $\qquad\qquad$ ◻

Problem 4.29 Let B be a cube. Find the orders of $S_R(B)$ and $S(B)$ respectively.

Solution. We suppose B is inscribed in the sphere of radius 1 centred at O. For ease of notation let us number the vertices of the cube as 1, 2, 3, 4, 5, 6, 7, 8 (Figure 4.7). We shall calculate $|S_R(B)|$ by counting the possible axes of rotation of B.

We have rotational symmetry about an axis joining the midpoints of pairs of opposite sides through angles $\frac{1}{2}\pi$, π, $\frac{3}{2}\pi$, 2π. For example, the rotation, in Figure 4.7, about the axis joining the midpoints of the sides 1, 2, 3, 4 to 5, 6, 7, 8 through an angle $\frac{1}{2}\pi$ is represented by the permutation

$$\begin{pmatrix} 1 & 2 & 3 & 4 & 5 & 6 & 7 & 8 \\ 2 & 3 & 4 & 1 & 6 & 7 & 8 & 5 \end{pmatrix} = (1\ 2\ 3\ 4)(5\ 6\ 7\ 8).$$

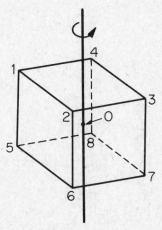

Figure 4.7

There are three axes of this type each with three nonidentity rotations and so there are in all nine such nonidentity rotations.

We have rotational symmetry about an axis joining opposite vertices through angles $\frac{2}{3}\pi$, $\frac{4}{3}\pi$, 2π. For example the rotation in Figure 4.8, about the axis joining 2 and 8 through an angle $\frac{2}{3}\pi$ is represented by the permutation

$$\begin{pmatrix} 1 & 2 & 3 & 4 & 5 & 6 & 7 & 8 \\ 6 & 2 & 1 & 5 & 7 & 3 & 4 & 8 \end{pmatrix} = (1\ 6\ 3)(4\ 5\ 7).$$

85

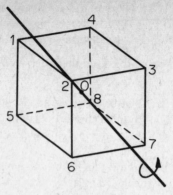

Figure 4.8

There are four axes of this type, each with two nonidentity rotations and so there are in all eight such nonidentity rotations.

Finally we have rotational symmetry about an axis joining the midpoints of diagonally opposite sides through π, 2π. For example the rotation, in Figure 4.9, about the axis joining the midpoints of 1,2 and 7,8 through an angle π is represented by the permutation

$$\begin{pmatrix} 1 & 2 & 3 & 4 & 5 & 6 & 7 & 8 \\ 2 & 1 & 5 & 6 & 3 & 4 & 8 & 7 \end{pmatrix} = (1\ 2)(3\ 5)(4\ 6)(7\ 8).$$

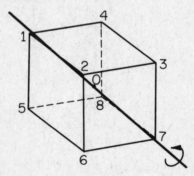

Figure 4.9

There are six axes of this type, each with one nonidentity rotation, giving six such nonidentity rotations.

We deduce that

$$|S_R(B)| = 9+8+6+1 = 24.$$

The cube is also symmetrical with regard to a reflection in a plane through

O parallel to a side of the cube. Thus if the plane is parallel to 1, 2, 3, 4 we have the reflection represented by

$$\begin{pmatrix} 1 & 2 & 3 & 4 & 5 & 6 & 7 & 8 \\ 5 & 6 & 7 & 8 & 1 & 2 & 3 & 4 \end{pmatrix} = (1\ 5)\,(2\ 6)\,(3\ 7)\,(4\ 8).$$

There are three such reflections but it may be shown, and the reader may verify experimentally with a model of a cube, that every symmetry of the cube is either a rotation or a rotation followed by the particular reflection $(1\ 5)\,(2\ 6)\,(3\ 7)\,(4\ 8)$. Hence $|S(B)| = 2|S_R(B)| = 48$.

The solid figure C whose six vertices are the centroids of the faces of the cube B is a regular octahedron, an eight-sided figure with equilateral faces. Every symmetry of C is a symmetry of B and conversely. In consequence the group $S_R(B)$ of rotational symmetries of the cube is sometimes called the *octahedral group*. It can be shown that $S_R(B)$ is isomorphic to the symmetric group S_4. $\qquad\square$

The next problem is entirely an exercise on elementary number theory and may be omitted on a first reading.

Problem 4.30 Let N be a strictly positive integer and let $n_1, n_2, \ldots, n_k$ be positive integers dividing N such that $2 \leqslant n_1 \leqslant n_2 \leqslant \ldots \leqslant n_k \leqslant N$. Suppose

$$\sum_{i=1}^{k} \left(1 - \frac{1}{n_i}\right) = 2\left(1 - \frac{1}{N}\right).$$

Prove that the following exhaust the possibilities:

$$k = 2 : n_1 = n_2 = N$$
$$k = 3 : n_1 = n_2 = 2, \qquad n_3 = \tfrac{1}{2}N$$
$$k = 3 : n_1 = 2, \qquad n_2 = n_3 = 3, \qquad N = 12$$
$$k = 3 : n_1 = 2, \qquad n_2 = 3, \qquad n_3 = 4, \qquad N = 24$$
$$k = 3 : n_1 = 2, \qquad n_2 = 3, \qquad n_3 = 5, \qquad N = 60.$$

Solution. A quick verification shows us that the listed possibilities do satisfy the given conditions. We have to show that these possibilities are exhaustive.

If $k = 1$ we have $1 \leqslant 2\left(1 - \dfrac{1}{N}\right) = 1 - \dfrac{1}{k_1}$ which is false. If $k \geqslant 4$ we have

$$2 > 2\left(1 - \frac{1}{N}\right) = \sum_{i=1}^{k}\left(1 - \frac{1}{n_i}\right) \geqslant k\left(1 - \frac{1}{n_1}\right) \geqslant k\left(1 - \frac{1}{2}\right) \geqslant 2 \quad \text{which is}$$

false. Thus we must have $k = 2, 3$.

87

Suppose now $k = 2$. Then

$$\frac{1}{n_1} + \frac{1}{n_2} = \frac{2}{N}.$$

If $n_2 < N$ we have $n_1 \leqslant n_2 < N$ and $\frac{1}{n_1} + \frac{1}{n_2} > \frac{2}{N}$ which is a contradiction.
Thus $n_2 = N$ and so $n_1 = N$.

Suppose $k = 3$. Then

$$\frac{1}{n_1} + \frac{1}{n_2} + \frac{1}{n_3} = 1 + \frac{2}{N}.$$

If $n_1 \geqslant 3$, since $n_1 \leqslant n_2 \leqslant n_3 \leqslant N$ we have

$$1 + \frac{2}{N} \leqslant \frac{1}{3} + \frac{1}{3} + \frac{1}{3} = 1$$

and so $n_1 = 2$ and hence

$$\frac{1}{n_2} + \frac{1}{n_3} = \frac{1}{2} + \frac{2}{N}.$$

If $n_2 = 2$ then $n_3 = \frac{1}{2}N$. If $n_2 \geqslant 4$ we have

$$\frac{1}{2} = \frac{1}{4} + \frac{1}{4} \geqslant \frac{1}{n_2} + \frac{1}{n_3} = \frac{1}{2} + \frac{2}{N}$$

which is false. Thus $n_2 = 3$ and we have

$$\frac{1}{n_3} = \frac{1}{6} + \frac{2}{N}.$$

We deduce easily that $n_3 < 6$ and so $n_3 = 3, 4, 5$ giving $N = 12, 24, 60$ respectively. $\square$

Problem 4.31 Find the orders of all finite groups of rotations whose axes pass through a fixed point.

Solution. Let the fixed point be the origin O and let G be the group having order N. G is a permutation group on the poles of the nonidentity rotations (Problem 4.27). Each pole belongs to an orbit under G and we may suppose there are k orbits. Let P_i be a pole of the ith orbit $G(P_i)$, then the ith orbit $G(P_i)$ has $|G : \mathrm{Stab}_G(P_i)|$ poles $(i = 1, 2, \ldots, k)$ (Problem 4.20). Thus we have $\sum_{i=1}^{k} |G : \mathrm{Stab}_G(P_i)|$ poles but what we require to devise is a method of counting poles that takes into account the fact that a given pole may be the pole of more than one rotation of G. We proceed as follows. Let X be the set of ordered pairs of the form (r, P) where r is a nonidentity rotation in G and P is a pole of r. Now each of the $N-1$ non-identity rotations has 2 distinct poles and so X consists of $2(N-1)$ ordered pairs. But we have an alternative, more complicated, method of counting the elements of X. The number of nonidentity rotations having P as a pole is the number of nonidentity rotations fixing P, the latter number is $\left|\mathrm{Stab}_G(P)\right|-1$. Thus the number of ordered pairs (r, P) for which $P \in G(P_i)$ is

$$\sum_{P \in G(P_i)} \left[\left|\mathrm{Stab}_G(P)\right|-1\right].$$

But for any $P \in G(P_i)$ we have $G(P) = G(P_i)$ and $\left|\mathrm{Stab}_G(P)\right| = \left|\mathrm{Stab}_G(P_i)\right|$ (Problem 4.20). Thus the number of order pairs (r, P) for which $P \in G(P_i)$ is

$$\sum_{P \in G(P_i)} \left[\left|\mathrm{Stab}_G(P_i)\right|-1\right] = \left|G : \mathrm{Stab}\,(P_i)\right|\left[\left|\mathrm{Stab}_G(P_i)\right|-1\right].$$

Hence, letting $\left|\mathrm{Stab}_G(P_i)\right| = n_i$ $(i = 1, 2, \ldots, k)$ the number of ordered pairs in X is

$$\sum_{i=1}^{k} \left|G : \mathrm{Stab}\,(P_i)\right|\left[\left|\mathrm{Stab}_G(P_i)\right|-1\right] = \sum_{i=1}^{k} \frac{N}{n_i}(n_i-1)$$

$$= N \sum_{i=1}^{k} \left(1-\frac{1}{n_i}\right).$$

We have therefore

$$\sum_{i=1}^{k} N\left(1-\frac{1}{n_i}\right) = 2(N-1)$$

or

$$\sum_{i=1}^{k} \left(1-\frac{1}{n_i}\right) = 2\left(1-\frac{1}{N}\right).$$

By Problem 4.30 $k = 2, 3$. If $k = 2$ then $n_1 = n_2 = N$ and there are two orbits each with one pole; in other words the nonidentity rotations all have the same poles. If $k = 3$ then there are several possibilities; if $n_1 = n_2 = 2, n_3 = \frac{1}{2}N$ there are 3 orbits having $\frac{1}{2}N, \frac{1}{2}N, 2$ poles respectively and if $n_1 = 2, n_2 = 3$ we have $N = 12, 24$ or 60.

The rotation groups of orders 12 and 24 have already been given, they are the tetrahedral and octahedral respectively (there being no other rotation groups of these orders). There is one rotation group of order 60, the so-called *icosahedral group*; this group is the group of rotational symmetries of the regular icosahedron which is a regular figure of twenty equilateral faces. The regular dodecahedron, a figure of twelve regular pentagonal faces, is obtained from the regular icosahedron by taking as vertices the centroids of the faces of the icosahedron and so the dodeca-hedron also has the icosahedral group as its group of rotational symmetries. It can be shown that the icosahedral group is isomorphic to the alternating group A_5.

EXERCISES

1. Prove that the centraliser of the permutation $\begin{pmatrix} 1 & 2 & 3 & 4 \\ 2 & 1 & 4 & 3 \end{pmatrix}$ in S_4 has order 8.

2. Prove that the permutation $(1 \quad 2 \quad 3)(2 \quad 4 \quad 1)(3 \quad 4 \quad 5)(2 \quad 4 \quad 6)$ of S_6 is the product of a transposition and a cycle of length 4.

3. The real polynomial $f(x_1, x_2, x_3)$ has degree 2 and is symmetric over S_3. Prove that $f(x_1, x_2, x_3)$ has the form $a + b(x_1 + x_2 + x_3) + c(x_1 x_2 + x_2 x_3 + x_3 x_1) + d(x_1^2 + x_2^2 + x_3^2)$ for suitable constants $a, b, c, d \in \mathbb{R}$.

4. Let G be a permutation group on a set X and let Y be an orbit under G. Prove that G acts as a transitive permutation group on Y.

5. The symmetric group S_5 acts on the polynomial $p = x_1^3 + x_2^3$ as a group of permutations. Prove that $\text{Stab}_G(p)$ is isomorphic to the direct product $C \times S_3$ where C is a cyclic group of order 2 and S_3 is the symmetric group of order 6.

6. Prove that the symmetry group of a rectangle which is *not* a square is isomorphic to the Klein four-group. Show that the Klein four-group is generated abstractly by a, b with the relations $a^2 = e, b^2 = e, ab = ba$.

7. Prove that the symmetry group $S(U)$ of a square U has generators a, b with the relations $a^4 = e, b^2 = e, b^{-1}ab = b^{-1}$. Prove that $S(U)$ has order 8 and that the centre of $S(U)$ has order 2.

8. Let G be a finite group of rotations, the nonidentity rotations all having the same axis. Prove that there is a least, nonzero, angle θ such that every rotation is through an angle which is an integral multiple of θ. Deduce that G is cyclic.

9. Considered as a figure in $\mathbb{R}^3$ let L_n be the regular polygon of n sides with vertices at the n points having cylindrical polar coordinates (r, θ, z) where $r = 1$, $\theta = 2k\pi/n$, $z = 0$ ($k = 0, 1, \ldots, n-1$). Prove that the symmetry group of rotations (in $\mathbb{R}^3$) of L_n is the dihedral group D_n of order $2n$. Show that the $2n+2$ poles of the rotations fall into three orbits which

are given as follows: if n is odd the orbits are

$$\{(r, \theta, z) : r = 1, \quad \theta = 2k\pi/n, \quad z = 0; \quad k = 0, 1, \ldots, n\}$$
$$\{(r, \theta, z) : r = 1, \quad \theta = \pi + 2k\pi/n, \quad z = 0, \quad k = 0, 1, \ldots, n\}$$
$$\{(r, \theta, z) : r = 0, \quad \theta = 0, \quad z = \pm 1\}$$

and if n is even the orbits are

$$\{(r, \theta, z) : r = 1, \quad \theta = 2k\pi/n, \quad z = 0; \quad k = 0, 1, \ldots, n-1\}$$
$$\{(r, \theta, z) : r = 1, \quad \theta = (2k+1)\pi/n, \quad z = 0; \quad k = 0, 1, \ldots, n-1\}$$
$$\{(r, \theta, z) : r = 0, \quad \theta = 0, \quad z = \pm 1\}.$$

Chapter 5

Further Group Theory

5.1 p-Groups, Sylow subgroups We begin with a problem that extends some of our earlier work on homomorphisms.

Problem 5.1 Let G, H be groups and let $f: G \to H$ be an epimorphism with kernel K. Let N be a subgroup of H. Prove that there is a unique subgroup M of G containing K such that $f(M) = N$. If N is normal in H prove that M is normal in G.

Solution. We let $M = \{x : x \in G, \quad f(x) \in N\}$. Then M is a subgroup of G since, for x, $y \in M$, we have $f(xy) = f(x)f(y) \in N$ and $f(x^{-1}) = [f(x)]^{-1} \in N$. By definition $K \subseteq M$ and if L is a subgroup of G such that that $f(L) \subseteq N$ then $L \subseteq M$; thus M is unique. Suppose N is normal in H. Then, for $m \in M$, $g \in G$, $f(g^{-1}mg) = f(g^{-1})f(m)f(g) = [f(g)]^{-1}f(m)f(g) \in N$ and so M is normal in G.

In the particular case that $H = G/K$ and that f is the natural homomorphism $f(x) = Kx \, (x \in G)$ we see that $M = \{x : x \in G, \quad Kx \in N\} = \bigcup_{Kx \in N} Kx$. Thus we have the convenient result that $N = M/K$ or, in other words, any subgroup N of G/K may be written as M/K for a (unique) subgroup M of G. $\square$

Our next problem is a useful 'counting' result for finite groups.

Problem 5.2 Let G be a finite group with centre $Z(G)$. Let $\mathscr{C}_1, \mathscr{C}_2, \ldots, \mathscr{C}_r$ be those conjugacy classes of G that contain two or more elements of G. Let $x_i \in \mathscr{C}_i$ and let $C_G(x_i)$ be the centraliser of x_i in $G \, (i = 1, 2, \ldots, r)$. Prove that

$$|G| = |Z(G)| + \sum_{i=1}^{r} |G : C_G(x_i)|. \tag{5.1}$$

Solution. Since conjugacy is an equivalence relation (Problem 2.29), G is the disjoint union of its equivalence classes. But $Z(G)$ is precisely the set of self-conjugate elements of G and so $G = Z(G) \cup \mathscr{C}_1 \cup \mathscr{C}_2 \cup \ldots \cup \mathscr{C}_r$ is a disjoint union. Hence (Problem 2.53) we have

$$|G| = |Z(G)| + \sum_{i=1}^{r} |G : G_G(x_i)| \qquad \square$$

Problem 5.3 Let G be a group of order $p^a \, (a \geqslant 1)$ where p is a prime. Prove that the centre $Z(G)$ of G is nontrivial.

Solution. In the notation of the previous problem we have $p^a = |G| = |Z(G)| + \sum_{i=1}^{r} |G : C_G(x_i)|$. Since $|G : C_G(x_i)|$ divides p^a and $|G : C_G(x_i)| > 1$ we have $|G : C_G(x_i)| = p^{a_i}$ where $a_i \geqslant 1$ $(i = 1, 2, ..., r)$. Thus, certainly, p divides $\sum_{i=1}^{r} |G : C_G(x_i)|$ and so p divides $|Z(G)|$. This establishes the assertion. $\square$

Problem 5.4 Prove that a group G of order p^2, where p is a prime, is Abelian.

Solution. (cf. Problem 3.7.) Since the centre $Z(G)$ of G has order p or p^2 we require to show that $|Z(G)| = p^2$. Suppose $|Z(G)| = p$. Then $|G/Z(G)| = p$ and so $G/Z(G)$ is cyclic, being generated by any coset except $Z(G)$ itself. Hence (Problem 2.69) G is Abelian and this contradicts the supposition that $|Z(G)| = p$. Thus G is Abelian.

We remark that groups of order p^3 are not necesarily Abelian; roughly speaking, the complexity of the group structure increases very much with the power p^a of p. $\square$

Problem 5.5 Prove that there are two non-Abelian nonisomorphic groups of order 8.

Solution. We have already encountered the quaternion group (Problem 2.68) and the dihedral group D_4 (Problem 4.26, Exercise 7 of Chapter 4). In the notation of Problem 2.68 the quaternion group is generated by b, d with the relations $b^4 = e$, $d^2 = b^2$, $d^{-1}bd = b^{-1}$ and we observe that there is exactly one element of order 2, namely $a = b^2$. In the notation of Exercise 7 of Chapter 4, the dihedral group D_4 is generated by a, b with the relations $a^4 = e, b^2 = e, b^{-1}ab = a^{-1}$ and we observe that there are two elements of order 2, namely a^2, b. The observations on the orders of the elements establish that the groups are not isomorphic.

We notice that the groups are similar; if G denotes either of the two groups then G has a centre $Z(G)$ of order 2 and $G/Z(G)$ is isomorphic to the Klein four-group. In general it is not an easy problem to determine whether two, apparently similar, groups are isomorphic or not. $\square$

Problem 5.6 Let G be a group of order p^a $(a \geqslant 1)$ where p is a prime. Prove that G has a normal subgroup of index p.

Solution. The result is trivially true if $a = 1$ since the subgroup consisting of the identity has index p in G. Suppose $a > 1$, we employ an induction argument assuming as our hypothesis that the result is true for any group of order p^b, $b < a$. We know that G has a nontrivial centre

$Z(G)$ (Problem 5.3). Let $x \in Z(G)$, $x \neq e$. Then x generates a cyclic group of order dividing p^a and so a suitable power of x generates a cyclic subgroup K of order p (Problem 2.42). Since $K \subseteq Z(G)$, K is normal in G and we consider G/K. By our assumption G/K contains a normal subgroup N such that $|G/K : N| = p$. By Problem 5.1, $N = M/K$ for some normal subgroup M of G and then

$$|G : M| = \frac{|G|}{|M|} = \frac{|G|}{|K|} \cdot \frac{|K|}{|M|} = \frac{|G/K|}{|M/K|} = |G/K : M/K| = |G/K : N| = p.$$

$\square$ (5.2)

Problem 5.7 Let G be a group of order $p^a m$ where p is a prime not dividing m. Prove that G has a subgroup of order p^a.

Solution. We remark first that if G is Abelian, the result is already known to us (Problem 3.28). Suppose G is non-Abelian, we make the induction hypothesis that the assertion is true for all groups of strictly lower order than G. In the notation of Problem 5.2 we have $p^a m = |G| = |Z(G)| + \sum_{i=1}^{r} |G : C_G(x_i)|$. If p divides $|G : C_G(x_j)|$ $(j = 1, 2, \ldots, r)$ then p divides $|Z(G)|$ and so $Z(G)$ has a p-Sylow subgroup Q of order p^b (say) where $b \leqslant a$. Since Q is a central subgroup of G, G/Q exists and $|G/Q| = p^{a-b} m$. By our hypothesis G/Q has a subgroup of order p^{a-b} and we may let P/Q be this subgroup where P is a subgroup of G (Problem 5.1). Then $|P| = |P/Q||Q| = p^{a-b} \, p^b = p^a$. If, now, p does not divide $|G : C_G(x_j)|$ $(j = 1, 2, \ldots, r)$ there exists k such that p does not divide $|G : C_G(x_k)|$. But $p^a m = |C_G(x_k)||G : C_G(x_k)|$ and so p^a divides $|C_G(x_k)|$. Since $|C_G(x_k)| < |G|$ we infer that $C_G(x_k)$ has a subgroup of order p^a. We have now completed the induction argument.

In accordance with our definitions for Abelian groups, a subgroup of G of order p^a, where $|G| = p^a m$ and p does not divide m, is called a *p-Sylow subgroup of G*. A group all of whose elements have orders which are powers of a given prime is called a *p-group*, it follows from this result that a finite group is a p-group if and only if the order of the group is a power of p. $\square$

Problem 5.8 Find the orders of the Sylow subgroups of a group of order 1 064 800.

Solution. Since $1\,064\,800 = 2^5 \times 5^2 \times 11^3$ the Sylow subgroups have orders 32, 25, 1331.

Problem 5.9 Find the three 2-Sylow subgroups of S_4.

Solution. A_4 and S_4 have a normal subgroup H of order 4, namely

94

$H = \{(1), (1\ 2)(3\ 4), (1\ 3)(2\ 4), (1\ 4)(2\ 3)\}$. H has index 2 in each of the three 2-Sylow subgroups P_1, P_2, P_3 where $P_1 = H \cup H(1\ 2), P_2 = H \cup H(1\ 3)$, $P_3 = H \cup H(1\ 4)$. It is fairly easy to verify that P_1, P_2, P_3 are subgroups of order $8 = 2^3$.

It may be shown that there are no other 2-Sylow subgroups of S_4. $\square$

Problem 5.10 Let P be a normal p-Sylow subgroup of the finite group G and let Q be a p-subgroup of G. Prove that $Q \subseteq P$.

Solution. Since Q is normal in G, PQ is a subgroup of G (Problem 2.45) and $|PQ| = \dfrac{|P||Q|}{|P \cap Q|} = |P||Q : (P \cap Q)|$ (Exercise 21 of Chapter 2). If $P \cap Q$ is a proper subgroup of Q then $|Q : (P \cap Q)| = p^b$ where $b \geqslant 1$. Then $|PQ| = |P|p^b$ and, on the other hand, $|PQ|$ divides $|G|$ and the highest power of p dividing $|G|$ is $|P|$. Thus $P \cap Q$ is not a proper subgroup of Q and so $Q = P \cap Q$ or, equivalently, $Q \subseteq P$ (Problem 1.7). $\square$

Problem 5.11 Let P be a p-Sylow subgroup of the finite group G and let $N_G(P)$ be the normaliser of P in G. Let Q be a p-subgroup of $N_G(P)$. Prove that $Q \subseteq P$.

Solution. By assumption P is a normal subgroup of $N_G(P)$ and so, by the previous problem, $Q \subseteq P$.

We observe that if, in particular, P is a normal Sylow subgroup of a finite group G then P is the unique p-Sylow subgroup of G. A finite group, all of whose subgroups are normal, is called *nilpotent*; a finite nilpotent group is the direct product of its Sylow subgroups. Any Abelian group is nilpotent. $\square$

Problem 5.12 Let G be a finite group and let H be a subgroup of G. Let $K_1, K_2, \ldots, K_n$ be n distinct subgroups of G forming a conjugacy class. If for $h \in H$ we define $h(K_i) = h^{-1}K_i h$ $(i = 1, 2, \ldots, n)$ prove that H is a permutation group on $\{K_1, K_2, \ldots, K_n\}$. If, further, under the action of H there are k orbits and if the subgroups are so numbered that K_1, $K_2, \ldots, K_k$ belong to distinct orbits, the remaining subgroups K_j being variously distributed into the orbits, prove that

$$n = \sum_{t=1}^{k} |H : \text{Stab}_H(K_t)|. \tag{5.3}$$

Solution. To confirm that H acts as a permutation group on $\{K_1, K_2, \ldots, K_n\}$ we observe that $h^{-1}K_i h$ is a conjugate of K_i and so belongs to $\{K_1, K_2, \ldots, K_n\}$ and that $h^{-1}K_i h = h^{-1}K_j h$ if and only if $i = j$. By Problem 4.20, $n = \sum_{t=1}^{k} |H : \text{Stab}_H(K_t)|$. $\square$

We use this result to establish an attractive and important result on the Sylow subgroups of a finite group.

Problem 5.13 Let G be a group and let P_1, P_2,..., P_n be n distinct p-Sylow subgroups of G such that $\{P_1, P_2,..., P_n\}$ is a conjugacy class of G. Prove that p divides $n-1$ and deduce that G has no other p-Sylow subgroups.

Solution. Let P be a p-Sylow subgroup of G. Then we may regard P as acting, as in Problem 5.12, as a permutation group on $\{P_1, P_2,..., P_n\}$. In the notation of Problem 5.12 and with a suitable renumbering we have

$$n = \sum_{t=1}^{k} |P : \text{Stab}_P(P_t)|. \tag{5.4}$$

But $\text{Stab}_P(P_t) = \{x : x \in P, \ x^{-1}P_t x = P_t\}$ and so $P = \text{Stab}_P(P_t)$ if and only if $x^{-1}P_t x = P_t$ for all $x \in P$. This implies that $P = \text{Stab}_P(P_t)$ if and only if P is a subgroup of the normaliser $N_G(P_t)$ of P_t in G. Thus $P = \text{Stab}_P(P_t)$ if and only if $P \subseteq N_G(P_t)$ or, equivalently (Problem 5.11), $P = P_t$. The conclusion is that $\text{Stab}_P(P_t)$ is a proper subgroup of P unless $P_t = P$. Let us suppose now that we choose $P = P_1$, then $|P : \text{Stab}_P(P_1)| = 1$ and p divides $|P : \text{Stab}_P(P_t)|$ $(t > 1)$. We infer that p divides $n-1$.

If G has p-Sylow subgroups other than $P_1, P_2,..., P_n$ let us suppose therefore that P is another p-Sylow subgroup. In this case we have that p divides $|P : \text{Stab}_P(P_t)|$ $(t = 1, 2,..., k)$ and so p divides n. As we have already shown that p divides $n-1$ we have a contradiction and so it must be false to suppse that G has p-Sylow subgroups other than $P_1, P_2,..., P_n$.

The result can be reformulated to say that the p-Sylow subgroups of a finite group G are all conjugate and that their number is of the form $1 + rp$ for some integer r ($\geqslant 0$) where, necessarily, $1 + rp$ divides $|G|$ (cf. Exercise 9 of chapter 2, Problem 2.53). ☐

Problem 5.14 Prove that a group of order 45 is Abelian.

Solution. Since $45 = 3^2 \times 5$, G has a 3-Sylow subgroup H of order 9 and a 5-Sylow subgroup K of order 5. Let H have n conjugates. Then $n = 1 + 3r$ ($r \geqslant 0$) and n divides 45 and so we conclude easily that, as the factors of 45 are 1, 3, 5, 9, 15, 45, $n = 1$ and hence H is normal in G. Similarly K is normal in G. We have $G = HK$ since $|HK| = |HK||H \cap K| = |H||K| = 45$ (Exercise 21 of Chapter 2) and thus G is isomorphic to the direct product $H \times K$. But H is Abelian (Problem 5.4) and K is cyclic and thus G is Abelian (Problem 3.2). ☐

We give two further problems illustrating some of the techniques used in handling Sylow subgroups.

Problem 5.15   Let G be a finite group and let P be a p-Sylow subgroup of G. Let $N_G(P)$ be the normaliser of P in G and let H be a subgroup of G such that $N_G(P) \subseteq H$. If $N_G(H)$ is the normaliser of H in G prove that $H = N_G(H)$.

Solution. Trivially $H \subseteq N_G(H)$. Let $x \in N_G(H)$, then $x^{-1}Px \subseteq x^{-1}Hx = H$. Thus $x^{-1}Px$ is a subgroup of H. We know that P is a p-Sylow subgroup of G and hence also is a p-Sylow subgroup of any subgroup of G containing P. Thus P and, by inference, also $x^{-1}Px$ are p-Sylow subgroups of H. But, by Problem 5.13 applied to H, there exists $h \in H$ such that $P = h^{-1}(x^{-1}Px)h = (xh)^{-1}P(xh)$. Hence $xh \in N_G(P)$ and so, as $N_G(P) \subseteq H$, we have $x = xhh^{-1} \in H$. Thus $N_G(H) = H$.   ☐

Problem 5.16   Let G be a finite group and let P be a p-Sylow subgroup of G. If N is a normal subgroup of G prove that $N \cap P$ is a p-Sylow subgroup of N.

Solution. Since P is a p-Sylow subgroup of the groups PN and G we may write $|PN| = |P|m_1$, $|G| = |P|m_1 m_2$ where m_1, m_2 are integers not divisible by p. Then (Exercise 21 of Chapter 2)

$$|N| = \frac{|P||N|}{|P|} = \frac{|PN||P \cap N|}{|P|} = |P \cap N|m_1. \tag{5.5}$$

Hence $P \cap N$, being a p-subgroup of N of highest possible order, is a p-Sylow subgroup of N.   ☐

5.2 Soluble groups   We begin with a result generalising Problem 3.14.

Problem 5.17   Let X be a nonempty subset of a group G. Let H be the subset of G consisting of all elements of the form $x_1^{a_1} x_2^{a_2} \ldots x_m^{a_m}$ where $x_i \in X, a_i = \pm 1 \ (i = 1, 2, \ldots, m; m = 1, 2, \ldots)$. Prove that H is a subgroup of G containing X.

Solution. Clearly $X \subseteq H$. If $x_1^{a_1} x_2^{a_2} \ldots x_m^{a_m}$ and $y_1^{b_1} y_2^{b_2} \ldots y_n^{b_n}$ are typical elements of H $(x_i, y_j \in X, a_i = \pm 1, b_j = \pm 1)$ then

$$x_1^{a_1} x_2^{a_2} \ldots x_m^{a_m} y_1^{b_1} y_2^{b_2} \ldots y_n^{n} \in H$$

and, by equation 2.31,

$$(x_1^{a_1} x_2^{a_1} \ldots x_m^{a_m})^{-1} = x_m^{-a_m} x_{m-1}^{-a_{m-1}} \ldots x_1^{-a_1} \in H.$$

Hence (Problem 2.31) H is a subgroup of G.

We remark that any subgroup of G containing X must contain products and inverses of elements of X and so contains H. H is then the least subgroup of G containing X and so is the intersection of all subgroups of G containing X. We say that H is the subgroup *generated* by X.   ☐

Problem 5.18 What is the subgroup of S_4 generated by $X = \{(1\ 2)(3\ 4),$ $(1\ 2\ 3)\}$?

Solution. Since $(1\ 2)(3\ 4)$ and $(1\ 2\ 3)$ are even permutations, X and the subgroup generated by X are subsets of A_4. On the other hand X generates A_4 for we have

$$(1\ 4)(2\ 3) = (1\ 2\ 3)(1\ 2)(3\ 4)(1\ 2\ 3)^{-1}$$
$$(1\ 3)(2\ 4) = (1\ 2\ 3)(1\ 4)(2\ 3)(1\ 2\ 3)^{-1}$$
$$(1\ 3\ 2) = (1\ 2\ 3)(1\ 2\ 3), \qquad (1\ 3\ 4) = /1\ 2\ 3)(1\ 2)(3\ 4)$$
$$(1\ 4\ 3) = (1\ 3\ 4)(1\ 3\ 4), \qquad (1\ 4\ 2) = (1\ 2\ 3)(1\ 4)(2\ 3)$$
$$(1\ 2\ 4) = (1\ 4\ 2)(1\ 4\ 2), \qquad (2\ 4\ 3) = (1\ 2\ 3)(1\ 3)(2\ 4)$$
$$(2\ 3\ 4) = (2\ 4\ 3)(2\ 4\ 3).$$
□

Problem 5.19 Let G be a group and let G' be the subgroup generated by all elements of the form $x^{-1}y^{-1}xy$ $(x, y \in G)$. Prove that G' is a normal subgroup of G and that G/G' is Abelian.

Solution. Let $z \in G'$, $g \in G$. To prove that G' is normal in G we have to show that $g^{-1}zg \in G'$. But, by the definition of G', $g^{-1}(z^{-1})^{-1}g(z^{-1}) \in G'$ and so $g^{-1}zg = g^{-1}(z^{-1})^{-1}g(z^{-1})z \in G'$. Thus G' is normal in G. Let now $x, y \in G$. Then

$$(xG')(yG') = xyG' = xy[(y^{-1}x^{-1}yx)G']$$
$$= (xyy^{-1}x^{-1}yx)G' = yxG' = (yG')(xG')$$

and so G/G' is Abelian.

An element of the form $x^{-1}y^{-1}xy$ $(x, y \in G)$ is called a *commutator* and G' is called the *(first) commutator* or *(first) derived* (sub)group of G. We observe that G' is trivial if and only if G is Abelian. We may repeat the process of forming derived groups. G'', the *second derived group* of G, is generated by commutators of elements in G' and the n*th derived group* $G^{(n)}$ is defined as $(G^{(n-1)})'$. If for some r, $G^{(r)}$ is trivial then G is said to be *soluble*. It may be shown that all groups of orders less than or equal to 60 are soluble except the alternating group A_5 (icosahedral group); it has been shown that all finite groups of odd order are soluble (in fact A_5 is simple—cf. Problem 2.72).
□

Problem 5.20 Let G be a group with derived group G'. Let H be a normal subgroup of G such that G/H is Abelian. Prove that $G' \subseteq H$.

Solution. Let $x, y \in G$. Since G/H is Abelian we have

$$x^{-1}y^{-1}xyH = (x^{-1}H)(y^{-1}H)(xH)(yH)$$
$$= (xH)^{-1}(yH)^{-1}(xH)(yH)$$
$$= (xH)^{-1}(xH)(yH)^{-1}(yH) = H$$

and so $x^{-1}y^{-1}xy \in H$. Thus $G' \subseteq H$.

We infer that G' is the intersection of all normal subgroups of G whose factor-groups are Abelian. $\quad\square$

Problem 5.21   Find the derived group S'_3 of the S_3.

Solution. The subgroup $A_3 = \{(1),(1\ 2\ 3),(1\ 3\ 2)\}$ is normal and the factor-group S_3/A_3 is cyclic of order 2. Thus (Problem 5.20) $S'_3 \subseteq A_3$. But $|A_3| = 3$ and so either $S'_3 = A_3$ or $S'_3 = \{(1)\}$. Since S_3 is non-Abelian $S'_3 = A_3$. $\quad\square$

Problem 5.22   Let G be a p-group of order p^3 with centre $Z(G)$ and derived group G'. If G is non-Abelian prove that $Z(G) = G'$ and that $|Z(G)| = p$.

Solution. We know that $|Z(G)| \geqslant p$ (Problem 5.3) and so, since G is non-Abelian, $|Z(G)| = p$ or p^2. If $|Z(G)| = p^2$ we have $|G/Z(G)| = p$ and so $G/Z(G)$ is cyclic. This implies G is Abelian (Problem 2.69). Hence $|Z(G)| = p$ and so $|G/Z(G)| = p^2$. By Problem 5.4 $G/Z(G)$ is Abelian and thus $G' \subseteq Z(G)$. Since $|Z(G)| = p$ either $G' = Z(G)$ or $G' = \{e\}$ and, by assumption, this latter alternative is not possible. $\quad\square$

Problem 5.23   Prove that a finite p-group is soluble.

Solution. Let G be a p-group of order p^a $(a \geqslant 1)$. Then G has a normal subgroup H of order p^{a-1} (Problem 5.6). Since G/H is cyclic, $G' \subseteq H$. Now we argue by induction and suppose the assertion is true for p-groups of orders p^b $(b < a)$. Then, in particular, H is soluble and so there exists n such that $H^{(n)}$, the nth derived group of H, is trivial. But $G' \subseteq H$ implies that $G'' = (G')' \subseteq H'$ and $G''' = (G'')' \subseteq (H')' = H''$ and so on. Thus $G^{(n+1)}$ is trivial and hence G is soluble. This completes the induction. $\quad\square$

Problem 5.24   Let $GL(2, \mathbb{R})$ be the general linear group of nonsingular 2×2 matrices over $\mathbb{R}$ and let $SL(2, \mathbb{R})$ be the special linear group of 2×2 matrices each having determinant 1. Prove that $SL(2, \mathbb{R})$ is the derived group of $GL(2, \mathbb{R})$.

Solution. We use the result that if $\mathbf{x}, \mathbf{y}$ are $n \times n$ matrices then $\det(\mathbf{xy}) = \det(\mathbf{x})\det(\mathbf{y})$. This relation asserts that mapping $\mathbf{x} \to \det(\mathbf{x})(\mathbf{x} \in GL(2, \mathbb{R}))$ is a homomorphism of $GL(2, \mathbb{R})$ into $\mathbb{R}^*$, the multiplicative group of nonzero real numbers. The kernel of this homomorphism is precisely $SL(2, \mathbb{R})$ which is therefore a normal subgroup of $GL(2, \mathbb{R})$. Furthermore, since $\mathbb{R}^*$ is Abelian, $[GL(2, \mathbb{R})]' \subseteq SL(2, \mathbb{R})$ (Problem 5.20).

The difficult part of our argument is to show that $SL(2, \mathbb{R}) \subseteq [GL(2, \mathbb{R})]'$ or, equivalently, to show that every 2×2 matrix of determinant 1 is a product of commutators. We consider some 'simple' matrices of determinant 1 namely

$$\begin{pmatrix} 1 & r \\ 0 & 1 \end{pmatrix}, \begin{pmatrix} 1 & 0 \\ s & 1 \end{pmatrix}, \begin{pmatrix} t & 0 \\ 0 & 1/t \end{pmatrix} \quad (r, s, t \in \mathbb{R}, \quad t \neq 0). \quad (5.6)$$

The following matrix calculations then show that these matrices are actually commutators.

$$\begin{pmatrix} 1 & -r \\ 0 & 1 \end{pmatrix} \begin{pmatrix} 1 & 0 \\ 0 & \frac{1}{2} \end{pmatrix} \begin{pmatrix} 1 & r \\ 0 & 1 \end{pmatrix} \begin{pmatrix} 1 & 0 \\ 0 & 2 \end{pmatrix} = \begin{pmatrix} 1 & r \\ 0 & 1 \end{pmatrix}$$

$$\begin{pmatrix} 1 & 0 \\ 0 & 2 \end{pmatrix} \begin{pmatrix} 1 & 0 \\ s & 1 \end{pmatrix} \begin{pmatrix} 1 & 0 \\ 0 & \frac{1}{2} \end{pmatrix} \begin{pmatrix} 1 & 0 \\ -s & 1 \end{pmatrix} = \begin{pmatrix} 1 & 0 \\ s & 1 \end{pmatrix}$$

$$\begin{pmatrix} 0 & 1 \\ 1 & 0 \end{pmatrix} \begin{pmatrix} t & 2t \\ 2t^2 & t^2 \end{pmatrix} \begin{pmatrix} 0 & 1 \\ 1 & 0 \end{pmatrix} \begin{pmatrix} -1/3t & 2/3t^2 \\ 2/3t & -1/3t^2 \end{pmatrix} = \begin{pmatrix} t & 0 \\ 0 & 1/t \end{pmatrix}.$$

But the matrices of (5.6) also generate $SL(2, \mathbb{R})$ since, if $\begin{pmatrix} a & b \\ c & d \end{pmatrix} \in SL(2, \mathbb{R})$ then $a, b, c, d \in \mathbb{R}$ and $ad - bc = 1$ and if $c \neq 0$ we have

$$\begin{pmatrix} a & b \\ c & d \end{pmatrix} = \begin{pmatrix} 1 & (a-1)/c \\ 0 & 1 \end{pmatrix} \begin{pmatrix} 1 & 0 \\ c & 1 \end{pmatrix} \begin{pmatrix} 1 & (d-1)/c \\ 0 & 1 \end{pmatrix}$$

and if $c = 0$ we have

$$\begin{pmatrix} a & b \\ 0 & d \end{pmatrix} = \begin{pmatrix} a & 0 \\ 0 & 1/a \end{pmatrix} \begin{pmatrix} 1 & b/a \\ 0 & 1 \end{pmatrix}.$$

Thus $SL(2, \mathbb{R})$ is generated by commutators and so $SL(2, \mathbb{R}) \subseteq [GL(2, \mathbb{R})]'$. This completes our argument. $\square$

EXERCISES

1. Let G, H be groups and let $f : G \to H$ be an epimorphism. If M is a normal subgroup of G prove that $f(M)$ is a normal subgroup of H.

2. Let G be a nontrivial p-group and let H be a proper subgroup of G. Prove that H is properly contained in its normaliser in G.

3. Prove that, up to isomorphism, there are exactly five groups of order 8.

4. Prove that the quaternion and dihedral groups of order 8 both have three subgroups of order 4.

5. In the notation of Problem 5.12 prove that $\text{Stab}_H(K_t) = H \cap N_G(K_t)$ where $N_G(K_t)$ is the normaliser in G of K_t.

6. Use the argument of Problem 5.13 to show that if Q is a p-subgroup of a finite group G then there exists a p-Sylow subgroup P of G such that $Q \subseteq P$.

7. Let G be a finite group and let K, P be a normal subgroup and a p-Sylow subgroup of G respectively. Prove that PK/K is a p-Sylow subgroup of G/K and that $G = KN_G(P \cap K)$. (cf. Problem 5.16).

8. Let G_i be groups with derived groups G_i' ($i = 1, 2$). Prove that $G_1 \times G_2$ has derived group $G_1' \times G_2'$.

9. Let G be a group and let G', K be the derived group and a normal subgroup of G respectively. Prove that $G'K/K$ is the derived group of G/K. Deduce that any homomorphic image of a soluble group is soluble. Show that a subgroup of a soluble group is soluble.

10. Let G be a group and let K be a normal subgroup of G such that K and G/K are soluble. Prove that G is soluble.

11. Prove that a nontrivial finite group is soluble if and only if every subgroup of order strictly greater than 1 has a normal subgroup of prime index.

Index

Abel, N. H., 27
Abelian group, 27, 55
Addition, 16, 18, 64
Alternating group, 75
Angle of rotation, 83
Associative Law, 11, 16
Axis of rotation, 83

Basis, 57
Bijection, 9
Bijective mapping, 9

$\mathbb{C}$, (Complex numbers), 6
Cayley, A., 76
Cayley's Theorem, 76
Central element, 30
Centraliser, 31
Centre, 30
Circle-notation, 9
Closed law of composition, 16
Commutator, 98
Commutator subgroups, 98
Commutative group, 20, 55
Commutative semigroup, 20
Commuting elements, 20
Composite mapping, 9
Conjugacy class, 30
Conjugate elements, 29, 30
Conjugate subgroups, 33
Coset (left), 36
Coset (right), 37
Coset decomposition, 37
Cyclic group, 27
Cycle, 70

Derived (sub-) group, 98
Descartes, R., 6
Dihedral group, 83
Direct product, 52, 54
Direct sum, 65
Disjoint cycles, 71
Disjoint union, 14
Domain, 7

Element, 1
Empty set, 1
Epimorphism, 42
Equivalence class, 14
Equivalence relation, 13
Even permutation, 75

Extension, 59

Factor-group, 45
Finite basis, 57
Finite order, 27
Finitely generated group, 56
Finitely generated free Abelian group, 57
First isomorphism theorem, 47
Free basis, 57
Function (cf. mapping), 7

General linear group, 31–32, 99–100
Generating element(s), 27, 56, 97
Generator, 27, 97
Group, 20
  Abelian, 27, 55
  Acting as permutation group, 76
  Alternating, 75
  Centre of, 30
  Commutative, 20, 55
  Coset of, 36–37
  Cyclic, 27
  Dihedral, 83
  Direct product, 52, 54
  Direct sum, 65
  Factor —, 45
  Finitely generated, 56
  Finitely generated free Abelian, 57
  General linear, 31–32, 99–100
  GL(2, $\mathbb{R}$), GL(n, $\mathbb{R}$), 31–32, 99–100
  Icosahedral, 89
  Klein four, 24
  Nilpotent, 95
  Octahedral, 87
  Order of, 27
  p-group, 61–62, 92–94
  p-Sylow subgroup, 62, 94
  Quaternion, 45
  Rank of, 58
  Simple, 48
  SL(2, $\mathbb{R}$), 99–100
  Soluble, 98
  Special linear, 99–100
  Symmetric, 67
  Symmetry, 77, 80
  Tetrahedral, 85
  Torsion-free, 56
  Trivial, 23